Journal of Social Issues

Fall 2000 Vol. 56, No. 3

spssi

Promoting Environmentalism

Issue Editors: Lynnette C. Zelezny and P. Wesley Schultz

NEW WAYS TO PROMOTE PROENVIRONMENTAL BEHAVIOR

ENVIRONMENTALISM AND PUBLIC POLICY

Issues in Progress
  The Impact of Welfare Reform
    Diana Zuckerman and Ariel Kalil
  Stigma: An Insider's Perspective
    Dapha Oyserman and Janet Swim

Journal of Social Issues, Vol. 56, No. 3, 2000, pp. 365–371

Promoting Environmentalism

Lynnette C. Zelezny*

California State University, Fresno

P. Wesley Schultz

California State University, San Marcos

This issue of the Journal of Social Issues *focuses on the psychology, sociology, and public policy of global environmental issues. The goals of the issue are (1) to describe the current state of the environment and environmentalism, (2) to present new theories and research on environmental attitudes and behaviors, and (3) to explore obstacles and ethical considerations in promoting proenvironmental action. The following brief introduction places this issue within a context of preceding* JSI *issues, provides a framework for the articles, and highlights several recurring themes that occur throughout the issue.*

As we begin the 21st century, humanity faces a number of pressing social issues. Looking back at recent issues of *JSI* reveals many of the issues on which psychologists have worked: sexual harassment, national and international group conflict, tobacco policy, health issues, youth violence, and prejudice and racism, to name just a few. Of all the social issues that face us in this millennium, the most daunting are environmental problems. In the last 100 years, humans have abused the planet on an unprecedented scale. The air, water, and land of the planet are becoming overused and polluted to the point where a large-scale environmental crisis is a real possibility. As Oskamp stated in a 1995 *JSI* article: "In this epoch of history, there is one danger that stands out as the most urgent and serious threat to the future of humanity—the threat of ecological disaster" (Oskamp, 1995, p. 217). Indeed, it is difficult (if not impossible) to find any part of the natural environment on this planet that has not been affected by human activity. If life on this planet is to

*Correspondence concerning this article should be addressed to Lynnette Zelezny, Department of Psychology, California State University, 5310 North Campus Dr., M/S PH 11, Fresno, CA 94740-8019 [e-mail: lynnette@csufresno.edu].

continue as we know it, humans must learn to balance growth and technological development with the resources and capacity of the planet.

That environmental problems are "social issues" is indisputable: They are caused by human behavior. What has been disputed is the role that psychology can play in addressing environmental issues. Environmental problems are large scale —they are created by the aggregate of human behavior—and solving these problems will require large-scale changes in behavior. Because of the level of change required to address environmental problems, what could psychology have to say about environmental problems? Indeed, this mismatch between psychology's level of analysis and the macro-level changes required to solve environmental problems has led some scholars to argue that psychology is "mute" about environmental problems (Kidner, 1994).

Psychology is not mute on this issue, however. Indeed, the changes that are required to solve our environmental crisis involve changes in individual behavior. Any policy, program, or intervention must produce change among individuals. An understanding of individual-level attitudes, motives, beliefs, intentions, or values will help to inform the development of such programs and also to explain why a particular program is, or is not, producing the desired changes. The interdisciplinary nature of environmental problems is clearly reflected in previous *JSI* issues that have focused on environment-related issues.

Preceding *JSI* Issues Devoted to Environmental Issues

Several previous issues of the *Journal* have contained articles addressing environmental problems. The first extended discussion of environmental problems was in a 1966 issue edited by Kates and Wohlwill titled "Man's Response to the Physical Environment." In the lead article (which was actually the 1966 SPSSI Presidential Address), Jerome Frank describes the direct effects of "man's reckless conquest of the environment" on human life and health. In this article, Frank argues that

> galloping technology has created or aggravated problems of unemployment, urbanization, racial and international tensions, war, overpopulation, and many others that have been the constant concern of members of SPSSI. . . . That this state of affairs is a pressing social issue seems self-evident, so it is appropriate to ask why it has aroused so little interest among social psychologists. The basic trouble may be that, in contrast to our other concerns such as war, poverty, or racial discrimination, this one has no focus and no villians. Ironically, the ills caused by technology are by-products of benevolent efforts to promote the general welfare. It is hard to get indignant over this, and indignation seems to be the initial goad to becoming concerned about a social issue. (pp. 1, 10)

Despite Frank's all-too-current lamentation, the articles in the 1966 issue did not focus on environmental problems. Almost without exception, the articles in the 1966 issue examined the ways in which the physical environment (both built and natural) affects human behavior. A similar focus can be found in a *JSI* issue edited by Cvetkovich and Earle (1992).

Although a few articles on environmental issues can be found earlier (i.e., Dunlap & Van Liere, 1977; Evans & Jacobs, 1981; Heberlein, 1972, 1977), the first issue of this journal devoted exclusively to environmental problems was Seligman and Becker's (1981) volume on "Energy Conservation." This issue was followed eight years later by an issue titled "Managing the Environment" by Seligman and Syme. Unlike preceding issues, these two volumes focused on the ways in which human behavior was affecting the natural environment and on strategies for changing human behavior.

This theme continued through the 1990s in two issues of *JSI* devoted to environmental issues. The first was Clayton and Opotow's 1994 issue on "Green Justice: Conceptions of Fairness and the Natural World." Articles in the this issue were unified in their application of "justice" to understand how people related to the natural environment and how to motivate people to act in a more proenvironmental manner. More recent was McKenzie-Mohr and Oskamp's 1995 issue on "Psychology and the Promotion of a Sustainable Future." This issue provided several cross-cutting frameworks for conceptualizing environmental problems and understanding the psychological forces that lead people to act in nonenvironmental ways. The issue also contained articles on applying psychological principles and theories to promote proenvironmental behavior.

The Articles in This Issue

This issue brings together the latest ideas, theories, innovations, and cutting-edge research from a field of recognized experts in psychology, sociology, and public policy to promote environmentalism. The multidisciplinary approach in this issue is unique; however, it represents an approach that we believe will be required to achieve significant environmental progress, that is, collective efforts across disciplines to effectively foster sustainable living using a broad knowledge base, the most creative ideas, and the most promising strategies.

"Promoting Environmentalism" advances the ideas presented in the previous *JSI* issues that tackled environmental concerns. Our focus is on promoting environmentalism and on psychological factors that lead people to act in proenvironmental ways. We use the term *environmentalism* broadly here to refer to *the processes associated with actions intended to lessen the impact of human behavior on the natural environment.* Within the broad framework of environmentalism, we include a variety of psychological constructs, including behavior, intentions, attitudes, beliefs, motives, and values. In addition, we include actions intended to lessen the impact of others on the natural environment, such as activism, public policy, and environmental justice. The issue contains 12 articles written by scholars from a variety of disciplines yet united in their focus on understanding and promoting environmentalism. We have organized these articles into six sections.

The first section is titled "Psychology of Promoting Environmentalism." Stuart Oskamp's article, "Psychological Contributions to Achieving an Ecologically Sustainable Future for Humanity," outlines and discusses the harmful impact human behavior has on the natural environment, including population growth, overconsumption, and underconservation. Oskamp also discusses obstacles to change, and offers some suggestions for overcoming these obstacles.

The second section is titled "New Environmental Theories." Two articles in this section attempt to piece together a broad, theoretical framework for integrating research findings on environmentalism. In the first article,"Empathizing With Nature: The Effects of Perspective Taking on Concern for Environmental Issues" Wesley Schultz argues that attitudes of environmental concern are rooted in a person's concept of self and the degree to which an individual perceives him- or herself to be an integral part of the natural environment. Two empirical studies are presented, suggesting that concern for the environment is related to empathy. In the second article in this section,"Toward a Coherent Theory of Environmentally Significant Behavior," Paul Stern provides a framework for understanding behavior that has significance for the environment. The article focuses on different types of environmentally significant behavior and discusses some of the theoretical approaches used by researchers to understand these actions. Stern distinguishes between the goals of environmentalism as a movement and environmentally relevant individual behavior. The author uses this distinction to articulate a broad framework for understanding proenvironmental behavior.

In the third section, "New Trends in Measuring Environmental Attitudes," Riley Dunlap, Kent Van Liere, Angela Mertig, and Robert Jones provide a revision to the widely used New Environmental Paradigm (NEP) scale that more comprehensively measures an ecological worldview. Their 15-item New Ecological Paradigm scale includes updated terminology and has higher internal reliability than the former NEP environmental attitudes scale. Their article reviews the range of existing measures of environmental concern and provides several suggestions for developing improved measures.

The fourth section, titled "New Ways of Thinking About Environmentalism," includes three articles. In the first, "Elaborating on Gender Differences in Environmentalism," Lynnette Zelezny highlights new evidence of gender differences in environmental attitudes and behaviors across age and countries. Females consistently report more concern for environmental issues and also more frequent proenvironmental behaviors, including political activism. Explanations for these differences are empirically examined using socialization theory. The second article, by Susan Clayton, examines "Models of Justice in the Environmental Debate." Clayton summarizes the models of justice used in public discourse about environmental issues. Data from two studies are presented indicating that responsibility to other species and to future generations, along with the rights of the environment, emerge as the two dominant considerations in addressing environmental conflicts.

Clayton advances the idea that environmental justice promotes environmental conflict resolution. The third article, by Susan Opotow and Leah Weiss, examines "Denial and the Process of Moral Exclusion in Environmental Conflict." Opotow and Weiss present the theory of moral exclusion and offer a conceptual analysis of the dynamics of denial and moral exclusion on environmentalism. They discuss the impact of this dynamic on environmental conflict resolution. The authors distinguish between three types of denial in environmental conflicts and the implications of denial for theory and practice.

The issue's fifth section, "New Ways to Promote Proenvironmental Behavior," offers four articles. The first, by Stephen Kaplan, examines "Human Nature and Environmentally Responsible Behavior." Kaplan suggests that the altruism-centered approach to environmentalism has been limiting; therefore, he proposes an alternative approach using insights from cognitive science and human evolution. Kaplan argues that behavior is determined by multiple motives; that is, the same action can be produced in two people using different motives. He suggests a Reasonable Person Model that, rather than imposing "good" motives in an attempt to produce "good" behavior, identifies reasons that individuals give for proenvironmental behavior. Conceptually, Kaplan's Reasonable Person Model posits that it may be easier to promote environmentally responsible behavior if we recognize human nature and circumstances that foster motivation. Kaplan's alternative approach focuses on reducing guilt and feelings of helplessness and supporting durable motivation and innovative problem solving in promoting environmentally responsible behavior. The second article, by Raymond De Young, deals with "Expanding and Evaluating Motives for Environmentally Responsible Behavior." In this article, De Young distinguishes between proenvironmental behavior motivated by incentives and that motivated by altruism. De Young suggests the need to broaden the range of motives in promoting environmentalism. Further, he examines a new motivational strategy for promoting responsible proenvironmental behavior using intrinsic satisfaction, which has impressive potential. He argues that promoting proenvironmental behavior depends on identifying motives that are strong, durable, generalizable, and culturally compatible with the individual. In the third article, "The Application of Persuasion Theory to the Development of Effective Proenvironmental Public Service Announcements," Renee Bator and Robert Cialdini discuss some of the important considerations in developing effective public service announcements. Bator and Cialdini provide specific guidelines, drawn from research on social influence and the Elaboration Likelihood Model, for developing effective public service announcements. The guidelines include measures of commitment and consistency, habits, and behavioral intentions. They discuss the importance of defining the optimal target audience for public service announcements, testing reactions using pilot messages, and ways to evaluate message effectiveness. The fourth article, by Doug McKenzie-Mohr, is titled "Promoting Sustainable Behavior: An Introduction to Community-Based Social

Marketing." In this article, McKenzie-Mohr laments the continued use of information-intensive campaigns to foster behavior change. He provides a four-step model for effective community-based social marketing. Case studies of backyard composting and water efficiency programs are presented to illustrate the process of social marketing.

The sixth and final section of the issue is titled "Environmentalism and Public Policy." In the issue's concluding article, "Environmental Justice: Grassroots Activism and Its Impact on Public Policy Decision Making," Robert Bullard and Glenn Johnson characterize the issue of environmental justice and document instances of policies and actions that are environmentally unjust. Bullard and Johnson document that people of color and low income unjustly face greater environmental and health risks than society at large. They discuss the recent momentum in grassroots activism, which has been strongly influenced by opposition to environmental racism and injustice, and the impact of environmental activism on community empowerment and policy.

Common Themes in This Issue

Despite the diversity among the contributing authors, the issue has several recurring themes. The first is a focus on integrating the research findings. The articles by Stern; Dunlap, Van Liere, Mertig, and Jones; Schultz; and Zelezny, Chua, and Aldrich synthesize past research to advance new conceptual and theoretical frameworks that are useful in promoting environmentalism. The second is a focus on power, explicit and implicit, and its relationship to environmental action and environmental justice. Broadly speaking, the articles by Oskamp; Zelezny, Chua, and Aldrich; Clayton; Opotow; and Bullard and Johnson examine disparate power, empowerment, and collective proenvironmental action. The third focus is on what prompts action. Ostensibly, the articles by Schultz; DeYoung; and Kaplan examine motives. Less ostensibly, the articles by Schultz; Zelezny, Chua, and Aldrich; and Clayton examine environmental values as guiding principles in environmental decision making and action. In addition, Clayton; Opotow and Weiss; and Bullard and Johnson examine justice as an impetus for environmental action. Finally, in the spirit of Kurt Lewin's action research, the fourth theme in this issue is applicability. Theory-driven applied research that is both scholarly and practical is exemplified in the articles by Oskamp; Bator and Cialdini; McKenzie-Mohr; and Bullard and Johnson. Effective and pragmatic strategies to promote environmentalism are clearly detailed. Hence, the focus is real-life application.

As we move into the 21st century, the impact of human behavior on the natural environment is becoming readily apparent. Resources are becoming less abundant, space is becoming more limited, and pollution of air, water, and land are beginning to have a direct impact on the inhabitants of the planet. The articles presented in this issue provide a variety of approaches for understanding, predicting, and changing environmentally significant behavior, all with the goal of promoting environmentalism.

References

Clayton, S., & Opotow, S. (Eds.) (1994). Green justice: Conceptions of fairness and the natural world. *Journal of Social Issues, 50*(3).

Cvetkovich, G., & Earle, T. (Eds.) (1992). Public responses to environmental hazards. *Journal of Social Issues, 48*(4).

Dunlap, R., & Van Liere, K. D. (1977). Land ethic and the golden rule: Comment on "The Land Ethic Realized" by Thomas A. Heberlein. *Journal of Social Issues, 33*(3), 200–206.

Evans, G. W., & Jacobs, S. V. (1981). Air pollution and human behavior. *Journal of Social Issues, 37*(1), 95–125.

Frank, J. D. (1966). Galloping technology, a new social disease [SPSSI Presidential Address]. *Journal of Social Issues, 22*(4), 1–14.

Heberlein, T. A. (1972). The land ethic realized: Some social psychological explanations for changing environmental attitudes. *Journal of Social Issues, 28*(4), 79–87.

Heberlein, T. A. (1977). Norm activation and environmental action: A rejoinder to R. E. Dunlap and K. D. Van Liere. *Journal of Social Issues, 33*(3), 207–210.

Kidner, D. W. (1994). Why psychology is mute about the environmental crisis. *Environmental Ethics, 16*, 359–372.

McKenzie-Mohr, D., & Oskamp, S. (Eds.) (1995). Psychology and the promotion of a sustainable future. *Journal of Social Issues, 51*(4).

Oskamp, S. (1995). Applying social psychology to avoid ecological disaster. *Journal of Social Issues, 51*(4), 217–239.

Seligman, C., & Becker, L. J. (Eds.) (1981). Energy conservation. *Journal of Social Issues, 37*(2).

Seligman, C., & Syme, G. (Eds.) (1989). Managing the environment. *Journal of Social Issues, 45*(1).

LYNNETTE ZELEZNY is an Assistant Professor of Psychology at California State University, Fresno, where she teaches courses in environmental psychology, applied social psychology, statistics, and research methodology. Her research interests are related to environmentalism, gender, and minority issues. Recent publications have focused on the effectiveness of environmental education, translational action research in minority mental health, and cross-cultural research on environmental attitudes and behaviors. She is the author of *Methods in Action* (1999, Wadsworth) and numerous publications on environmentalism.

P. WESLEY SCHULTZ is an Associate Professor of Psychology at California State University, San Marcos, where he teaches courses in psychology and statistics. His research interests are in the psychology of environmental issues and the application of psychological theory to help understand and solve social problems. Recent projects have focused on promoting recycling, cross-cultural research on environmental attitudes, environmental education, stereotypes, and prejudice. He is the coauthor of *Applied Social Psychology* (1998, Prentice-Hall) and *Social Psychology: An Applied Perspective* (2000, Prentice-Hall) and numerous research publications in the areas of social psychology and environmental issues.

Journal of Social Issues, Vol. 56, No. 3, 2000, pp. 373–390

Psychological Contributions to Achieving an Ecologically Sustainable Future for Humanity

Stuart Oskamp*

Claremont Graduate University

The most serious long-term threat facing the world is the danger that human actions are producing irreversible, harmful changes to the environmental conditions that support life on Earth. If this problem is not overcome, there may be no viable world for our descendants to inhabit. Because this threat is caused by human population growth, overconsumption, and lack of resource conservation, social scientists have a vital role in helping our world escape ecological disaster and approach a sustainable level of impact on the environment—one that can be maintained indefinitely. Enormous changes to human lifestyles and cultural practices may be required to reach this goal. This article discusses major obstacles to this goal, describes a variety of motivational approaches toward reaching it, and proposes that we should view the achievement of sustainable living patterns as a superordinate goal—a war against the common enemy of an uninhabitable world.

A central topic of this *Journal* issue is *sustainability*, that is, the urgent need for us to use the Earth's resources in ways that will allow human beings and other species to continue to exist acceptably on Earth in the future. In 1987, the Brundtland Report of the World Commission on Environment and Development (WCED) defined sustainable development as "meeting the needs of the present without compromising the ability of future generations to meet their own needs" (p. 363). Starkly stated, the issue is whether there will be a livable world for our descendants and other creatures to inhabit.

* Parts of this article are quoted or paraphrased from more extensive material in "A Sustainable Future for Humanity? How Can Psychology Help?" by S. Oskamp, 2000, *American Psychologist, 55*, 496–508 (copyright 2000 by the American Psychological Association; reprinted by permission). Later parts of the article add a psychological analysis of motivational issues relevant to sustainable living. Correspondence concerning this article should be addressed to Stuart Oskamp, Department of Psychology, Claremont Graduate University, Claremont, CA 91711 [e-mail: stuart.oskamp@cgu.edu].

Dangers to Earth's Environment

As a prelude to the articles that follow, I will briefly summarize some of the current drastic dangers to the Earth's environment (see Oskamp, 2000, for details). Most literate citizens are at least somewhat informed about them, and media coverage and popular awareness are increasing, but people typically are not aware of their potentially cataclysmic nature. It is important for all people to become environmentalists and to work toward reducing the damaging impacts of humans on the natural environment. Among the most serious dangers to the environment are

- Global warming due to the greenhouse effect. When oil, gas, coal, or wood are burned, the carbon dioxide (CO_2) that is produced mixes into the atmosphere. This CO_2 and other greenhouse gases absorb infrared radiation from Earth and thus reduce the amount of Earth's heat that is radiated into space, much as the glass roof in a greenhouse lets in warming sunlight but prevents warm air from escaping. According to the Intergovernmental Panel on Climate Change (IPCC), the amount of CO_2 in the atmosphere has been increasing very steadily and has reached levels unprecedented in geological history (IPCC, 1996). If this continues, the IPCC estimates that it will result in an average warming of the Earth's surface air temperature by about 3½ degrees Fahrenheit by the year 2100. This extra heat—even an average increase of 1 or 2 degrees—can change regional climates and disrupt agriculture worldwide. The polar regions are warming particularly fast, and a continuation of this trend will cause extensive melting of the polar icecaps, resulting in raised ocean levels and consequent flooding of huge low-lying coastal areas in many countries (Hileman, 1999; Schneider, 1997).

- Loss of much of the Earth's protective ozone layer due to release of chloroflourocarbons (CFCs). The extra ultraviolet radiation that penetrates the ozone layer causes damage to crops and skin cancer in humans (French, 1997; IPCC, 1996).

- Global climate change and great loss of biodiversity due to destruction of tropical and temperate rain forests (Bryant, Nielsen, & Tangley, 1997).

- Overfishing and exhaustion of all the world's oceanic fisheries and decreasing agricultural productivity due to many unsustainable practices. The world's grain production per person peaked in 1984, and the world's fish production per person peaked in 1989, and both have subsequently fallen by 7–8% (Brown, 1995, 1999). A likely future scenario is that increasing demand for food combined with a single summer's crop failure in one of the world's major agricultural nations will cause a dramatic escalation of world food prices and major famines in some nations.

- Acid rain, which damages forests and crops and also kills fish, plants, and other organisms in lakes and rivers (French, 1990; National Acid Precipitation Assessment Program, 1991).

- Toxic pollution of air and drinking water supplies. This is a worldwide problem (World Health Organization, 1992) resulting from humans overtaxing Earth's life-giving resources of air and water and its capacity for absorbing waste products.

- Genetic and hormonal damage and cancer due to exposure to dioxin and other toxic chemicals. A new and little-known example is research showing a nearly 50% decrease in average sperm count observed in men worldwide during the last 50 years, apparently due to the widespread use of chlorinated chemicals all over the world in those years (Colborn, Dumanoski, & Myers, 1996; Wright, 1996). Similarly, the dangerous carcinogen dioxin is now building up to alarming levels in the body tissues of most Americans (Schecter, 1994).

The Centrality of the Social Sciences

In thinking about environmental problems such as these, it is essential for us to realize that they are *not* solely technical problems, requiring simply engineering, physics, and chemistry for their solution. There is a crucial role for the social sciences in these problems because *they are all caused by human behavior, and they can all be reversed by human behavior.* Another key point is that *most of these problems are getting more serious each year,* so it is urgent that we do much more to reverse them (cf. Oskamp, 1995a). In fact, United Nations estimates predict that 20% of the world's population (nearly 2 billion people) will become "environmental refugees" by the year 2020 because of environmental damage in their areas, destruction of cropland, lack of water, and so on (George, 1993).

The United Nations Conference on Environment and Development, held in Rio de Janeiro, Brazil, in 1992, focused world attention on these problems and agreed on a plan of action for addressing them, called Agenda 21. Progress on these goals is being monitored by the UN Commission on Sustainable Development and by national agencies and citizen watchdog groups in many nations (Bartelmus, 1994). Progress has been sporadic, however, and slow at best, and the U.S. government has frequently played an obstructive role. For example, the 1997 international summit meeting in Kyoto, Japan, which was held to establish enforceable goals for nations to reduce the greenhouse gas emissions that are producing global warming, was impeded by U.S. government proposals that advocated minimal goals. As a result, it set only very weak and distant targets and established no enforcement mechanisms (Fishel, 1998; Lemonick, 1997).

Fortunately, opinion polls from many nations show that most people have high levels of concern for environmental problems. In the United States, pro-environmental attitudes hit an all-time high in the 1990s, and a large majority of people now call themselves "environmentalists" (Dunlap, Gallup, & Gallup, 1993; Kempton, Boster, & Hartley, 1995). Yet surprisingly, at this time of high public support for environmental preservation, many legislators in Congress and in some state legislatures are bent on reversing the environmentally protective legislation of the last 30 years. This attack is being supported by much of big business, such as oil and mining companies, cigarette manufacturers, and drug companies, with the claim that they are merely opposing unnecessary and wasteful government regulations. Unless voters are vigilant in registering their views, there are likely to be even more cutbacks in U.S. environmentally protective laws, regulations, and budgets.

There are three main sources of Earth's environmental problems: human over-population, overconsumption, and underconservation.

The Threat of Population Growth

The central source of the Earth's environmental problems is human population growth. Figure 1 shows the fantastic recent growth of the world's human population. For millions of years of human existence, the number of people on Earth remained small, eventually reaching 1 billion about 1800; then it took over 100

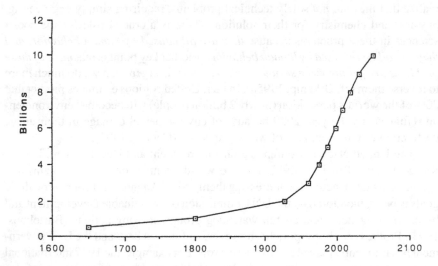

Fig. 1. World population growth since 1650
Source: The Population Reference Bureau, Inc.

years to reach 2 billion in 1930. It took only 30 years to reach 3 billion in 1960, and only 15 years to reach 4 billion in 1975. In 12 years the total reached 5 billion in 1987, and in another 12 years it hit 6 billion in 1999 (Brown & Flavin, 1999; David, 1994).

This geometric increase in human population is extremely dangerous for the Earth, and obviously this trend cannot continue much longer. The huge population increases have already brought miserable living conditions, malnutrition, and illness to about half of the people of the world. If we don't control human population growth voluntarily, it will eventually be controlled coercively. The death rate *will* catch up with the birth rate—either through starvation and famines, through diseases (e.g., AIDS, which is already ravaging Africa), or by wars and genocide (Hardin, 1993).

No one knows exactly how many people the Earth can support, but some environmental scientists have concluded that we may already have overshot Earth's long-term carrying capacity (Cohen, 1995; Hardin, 1993; Meadows, Meadows, & Randers, 1992). For instance, a study by Pimentel, Harman, Pacenza, Pacarsky, and Pimentel (1994) estimated that the Earth may be able to support only 3 billion people in perpetuity (about half the number of people alive today) and only 1–2 billion people if they are to live in "relative prosperity." That conclusion means that we are literally stealing vital life resources from our descendants.

The Trap of Overconsumption

The second main source of the Earth's environmental problems is over-consumption of natural resources, a widespread pattern that is like an addiction to unsustainable consumerism (cf. Clinebell, 1998). In particular, the affluent, industrialized nations are exhausting the natural resources of the poorer, developing nations by their overuse of energy and raw materials. The most extreme example is the United States. With only 5% of the Earth's population, the United States uses about 25% of the Earth's commercial energy, and uses it only about half as efficiently as Japan (Flavin & Dunn, 1999). Together, the industrial nations of the world have about 20% of the world's population, but they consume about 85% of the world's paper and automobiles (United Nations Development Programme, 1998). The level of use of the world's resources by a particular nation has been termed its "ecological footprint." Estimates indicate that increasing the rest of the world to the resource-use level of the United States and Canada would require the land and other natural resources of three Earths—an obviously impossible ambition (Wackernagel & Rees, 1996).

The result of worldwide overconsumption of the Earth's resources is that traditional supplies of many materials are being used up rapidly. A key example is petroleum: Its production has expanded greatly in the last 25 years, but known reserves have increased only slightly, with the result that world oil production will probably peak and begin to decline about 2010 (Campbell & Laherrere, 1998; Flavin & Dunn, 1999). Since much of the industrialized world runs on oil, this will have a dramatic impact on many aspects of life. It is true that modern nations have pioneered in developing synthetic materials such as plastics and chemicals to substitute for natural materials that are becoming scarce or costly. These synthetics, however, are mostly made from petroleum, further decreasing its reserves. Moreover, synthetic products made from organic chlorine compounds have been shown to have many dangerous ecological and health effects (e.g., Colborn & Clement, 1992; Misch, 1994), and plastics persist for centuries in trash and landfills without disintegrating.

The Tragedy of Underconservation

Paralleling the problem of overconsumption is the third main source of the Earth's environmental problems: underconservation of natural resources. The solution to worldwide overconsumption is twofold: reducing overall consumption and shifting to universal reuse of products (e.g., resale, remanufacturing, or sharing) and recycling of their constituent materials when the product's life is ended (G. Gardner & Sampat, 1999). These changes toward full conservation of raw materials would constitute a revolution in modern Western production and consumption practices, but there are already many successful examples of these needed conservation practices that can point the way for producers and consumers alike (cf. G. Gardner & Sampat, 1999).

The U.S. Environmental Protection Agency (1992) has popularized the slogan "reduce, reuse, recycle" to describe three key ways to decrease our overuse of natural resources. "Reduce" means attacking the problem of overconsumption by using less natural resources in the first place. "Reuse" and "recycle" refer to two ways of conserving raw materials by using them again. "Reuse" means using them again in the same form, whereas "recycling" means collecting manufactured products and changing their form (e.g., by crushing, shredding, or melting them down) for use in making new products.

Reuse

Many older Americans can remember when soft drink bottles (such as glass Coke bottles with their famous voluptuous shape) were returned to the bottler, sterilized, refilled with soft drinks, and resold. This reuse process saves the raw materials and the energy involved in making new containers for every batch of soft drink that is sold, and it also provides employment for individuals who collect and process the returned bottles (Shireman, 1993). It is still used in some places: for example, for refilling milk bottles in some European countries and soft drink bottles in many developing countries. Though this reuse process has been abandoned for most liquid consumable products in most developed nations, returning to it would save huge amounts of raw materials and avoid the environmental problems involved in recycling used materials or disposing of them in landfills or by incineration.

Recycling

Recycling has been adopted in the United States and many other developed countries as an easier way of ensuring that large amounts of natural resources are used again in productive ways rather than being dumped in landfills, where they often contribute to pollution problems. For instance, aluminum cans are melted down and used in making new aluminum products, many sorts of glass are collected and melted to make new glass containers, old tires are shredded and used as one ingredient in making new pavement, and recycled paper and rags are made into new paper products. In the case of some materials like aluminum, recycling can save enormous amounts of energy and pollution that are entailed in the original smelting of ore. However, there is always considerable waste resulting from original products that are not collected and recycled (around 10% in the case of aluminum cans—even in states where "bottle bills" provide cash payments for their return—and much higher percentages for most other products; Shireman, 1993). Also, the recycling process is less efficient than reuse because the raw material typically becomes degraded in recycling, and consequently the recycled products have lower quality than originally (e.g., recycled paper products have shorter fibers and therefore are less useful for some purposes).

The conclusion is clear that we have to stop destroying our environment (e.g., by polluting the air and water, dumping toxic and nuclear wastes, overdrawing water supplies, and overconsuming other natural resources). We have to transform the current destructive patterns of human behavior into ones that are sustainable over the centuries ahead.

Research Contributions to Sustainability

How are we going to meet this challenge and achieve sustainable patterns of living? As stated above, the social sciences are central to resolving problems that stem from patterns of human behavior, and they can make many contributions to the goal of sustainable lifestyles. As a reminder of their potential, consider these environmentally relevant areas that have been productively studied by social scientists (cf. Oskamp, 1995a).

- Population control and family planning programs have been disseminated widely (cf. Severy, 1993). A notable finding in this area is that increased economic opportunities for women constitute an important incentive to limiting family size (Abernethy, 1993).

- Energy conservation methods have been extensively studied (e.g., Katzev & Johnson, 1987). Research has shown that one-time investments, such as buying more efficient cars or installing home insulation, can save dramatically more energy than repeated minor actions like turning down thermostats or turning off lights (G. T. Gardner & Stern, 1996).

- Resource conservation and recycling are more recent areas of study. In the Los Angeles suburbs, my research group has found dramatically high rates of participation in local recycling programs (as high as 90%) but also a potential for recycling much greater amounts of materials (cf. Oskamp, 1995b).

- Avoiding global environmental changes is a crucial goal that is beginning to be studied (cf. Stern, Young, & Druckman, 1992). For long-term climate control, much more research is needed on ways of decreasing use of fossil fuels and of preserving both tropical and temperate rainforests.

- In accomplishing changes in environmental behavior, research has shown that the greatest success is likely to come from using a combination of different types of interventions (e.g., using educational interventions together with incentives, normative pressures, and/or removing barriers to change; Stern, this issue).

How Can We Communicate the Need for Living Sustainably?

Now let us consider how we can put across the message that we have to change to a sustainable lifestyle. Unfortunately, major lifestyle changes are often resisted, and this message is especially apt to be unpopular. The theme of using less resources suggests that we'll have to cut back on the comforts of life and "make do" with less. When someone brings an unwelcome message like this, the messenger is apt to be shot—or at best ignored! Consequently, political leaders are afraid to tell us; they promise growth and improvement rather than cutbacks or reductions. Similarly, many economists and business leaders believe in perpetual growth and try to persuade us that expansion can continue indefinitely. But remember the geometric curve of population growth: It can't continue!

Economists are so focused on growth that they count *losses*, such as resource destruction, as part of a nation's gross national product (GNP). For example, the GNP includes expenditures involved in using up all our oil reserves, cleaning up toxic dumps, and building nuclear missiles that we hope will never be used. These expenditures aren't productive of human welfare. Also, advocates of growth often don't consider what are called "externalities" when they compute the costs of products. For instance, the price of nuclear energy doesn't yet include the huge but inevitable "external" costs of eventually cleaning up nuclear wastes and decommissioning expired nuclear plants. In fact, we have not even figured out how to do these things.

Hence, the GNP is a misleading indicator of a society's welfare (Daly, 1996). A much better measure of a nation's genuine progress and productivity would be an "index of sustainable economic welfare" (ISEW). Such a measure should add the value of unpaid, voluntary work (e.g., housework, child care, volunteer programs) but subtract the costs of nonproductive expenditures such as pollution control, crime control, and depletion of nonrenewable resources (Cobb & Cobb, 1994). Developing such improved measures of sustainability is an important task for social scientists (e.g., Bartelmus, 1994).

How Much Behavioral Change Is Needed?

In the long run, our society will have to get by with using far less resources per person and will also have to reduce the number of people. In particular, because of the greenhouse effect, we must quickly and sharply reduce our use of fossil fuels and use them much more efficiently. The Environmental Protection Agency has estimated that avoiding global climate changes will require a 75% decrease in CO_2 emissions, and thus of fossil fuel use, continuing over many decades (Lashof & Tirpak, 1989). Because the United States is not only the largest user of energy, but also the most wasteful, our nation's behavior change is critical to the world.

The prophets of growth often propose that technology will solve this problem by using our resources much more efficiently (e.g., Ausubel, 1996). Indeed, technology *can* fill part of the gap—but only part. The widely publicized Brundtland Commission proposal for supposed "sustainable development," entitled *Our Common Future* (WCED, 1987), made the following forecast for the year 2050: a technology twice as efficient, for a world population twice as large as in 1987 (10 billion people), and a world economy 5–10 times greater than in 1987. The economics of this scenario sound attractive, but its environmental impact would be terrible!

A good way to analyze such scenarios for their environmental impact is the formula $I = PAT$ (Ehrlich & Ehrlich, 1991). It considers I (impact) to be a function of P (population size), A (affluence per person, which increases natural resource consumption), and reasonable estimates of T (technological improvements, which can decrease resource use per capita, if channeled wisely). Applying this formula to the Brundtland Report shows that its scenario is not a sustainable one: It would produce an environmental impact on the Earth two-and-a-half to five times the current level (Olson, 1995).

Achieving a sustainable society will require basic changes in everybody's behavior and in our basic values. As shown in several of the articles in this issue, environmental psychologists are engaged in debates about how best to achieve these changes and what kinds of variables are most important to emphasize (e.g., behaviors, values, attitudes, beliefs, incentives, norms, or barriers to behavior change).

In the realm of values, our nation has a particularly strong value of *human mastery over nature* (Kluckhohn & Strodtbeck, 1961), which is well illustrated by the injunction in the book of Genesis (I, 28): "be fruitful and multiply, and replenish the earth, and subdue it: and have dominion over . . . every living thing that moveth upon the earth." We need to change that value to one of *harmony with nature*, focusing on what will benefit all people and the whole world in the long run.

Obstacles to the Needed Changes

There are two major sources of opposition to the great changes needed for a sustainable world. The first source is national governments in many countries and multinational corporations, both of which profit hugely from consumption of resources, such as timber, oil, and minerals, and from processes that pollute the Earth, such as production of chemicals, plastics, and pesticides. In fact, multinational companies have become so powerful that they are laws unto themselves, often able to resist or overrule the edicts of national governments that have passed laws or regulations to safeguard or preserve natural resources. In these preservation efforts, local governments and community groups are often most actively involved because they are closer to the scene and they experience the dangers more directly. Excellent examples of grassroots environmental activism are presented in the article by Bullard and Johnson (this issue). Such local activists need as much help as possible from people everywhere in the world to fight against environmental destruction wherever it is occurring.

The second source of opposition to changes toward sustainability is individual people themselves, who very often resist major changes in their lifestyles. This resistance has many potential sources, including (a) inertia, which may postpone action until the environmental damage is irreversible, (b) the danger that appeals to fear will backfire and lead people to deny the environmental threats, (c) the belief that technological improvements alone can save the environment, and (d) opposition to the necessary lifestyle changes because they are perceived as requiring sacrifice and poverty.

The Issue of Motivation

This brings us to a central issue, the problem of motivation. How can we motivate ourselves and other people to make these vitally needed changes? De Young (this issue) has stressed that we should aim to appeal to motives that are both reliable and durable in their effects on behavior. Here I will briefly address each of the above sources of individual motivational resistance.

Inertia

Unfortunately, major lifestyle changes typically occur only in response to major crises (for instance, even the great behavioral changes produced by the AIDS epidemic have still not been enough to make most people's sexual behavior safe). Will we have to wait for a huge catastrophe that kills hundreds of millions of human beings before people everywhere become convinced of the need for drastic changes? Despite the drag of inertia, the motive of self-interest is a powerful one that we can appeal to in trying to create a healthier, less-polluted, sustainable world (cf. De Young, this issue).

In efforts to overcome inertia, publicity through the mass media is a crucial influence. Every available media avenue needs to be enlisted in the campaign to inform people, motivate them, and accomplish behavior change that will help save the environment. Yet at the same time, we need to reject the media's constant appeals for consumption, and that is a difficult combination to achieve.

Appeals to Fear

A basic difficulty in promoting behavioral change with respect to environmental issues is that the environmental problems we have been discussing stimulate fear, and social psychological research has clearly demonstrated the complications involved in using fear stimuli. Essentially, people don't like to think about fearful topics, and they frequently repress or deny such information. Clear examples can be seen in inattention to and denial of the dangers of global warming and the ozone hole. Despite the overwhelming scientific evidence to the contrary, there are state legislators and congressional representatives who are claiming that CFCs do not pose any environmental danger and are trying to get the United States to return to making and using them, thus violating the Montreal Protocol treaty, which banned production of CFCs. Similarly, many politicians have been objecting to even the weak and distant goals of the 1997 Kyoto climate conference as being damaging to U.S. industry and "progress," ignoring the drastic danger that global warming poses to living conditions on the entire planet (Broder, 1997).

Research studies on appeals to fear have shown that they are most likely to change people's behavior under two conditions: (1) if people are aware of clear steps they can take to protect themselves, and (2) if these steps are conveniently available (e.g., Leventhal, Meyer, & Nerenz, 1980). Unfortunately, because of the nature of environmental problems, neither of these conditions is easily met:

- Environmental problems are large, so people feel they can do little on their own.
- Environmental problems are long-term, so there are no immediate solutions.

Belief in Technology

As mentioned above, it is often proposed that technology will solve this problem by enabling us to use our resources much more efficiently (e.g., Ausubel, 1996; Simon, 1995), and most Americans have a great faith in progress stemming from technology. Though the Brundtland Commission proposal for supposed "sustainable development" does not actually represent a sustainable scenario, it would be wise to harness and use people's beliefs in technological progress, rather than trying to dispute them (see next section).

Dislike of Sacrifice and Poverty

It is apparently a universal human tendency to avoid losses and poverty and to dislike making sacrifices. Therefore it is apt to be counterproductive to describe sustainable living in terms of making sacrifices (cf. Kaplan, this issue).

An often-cited example of this tendency to avoid personal sacrifices is the so-called *tragedy of the commons* (Hardin, 1968). This term refers to people's typical pattern of using community-owned resources (like the environment) profligately, because they pursue their own short-term individual benefit and ignore the long-term negative consequences to their whole society. This outcome is even more likely when the long-term consequences can't easily be seen, such as the risk of cancer from chemical pollutants or of climate change from cutting down rainforests. However, offsetting this selfish human tendency, there is widespread research evidence showing that the condition of common natural resources (e.g., pastureland or crab fisheries) has prospered for long periods when managed by small community groups that developed and enforced a system of mutual norms and rules (e.g., Gardner & Stern, 1996, p. 30).

People have adaptation levels to their customary environmental conditions, and they generally display reactance against changes that they perceive as detrimental to them. Thus, the idea of sacrificing personal advantages or comforts in order to advance the general welfare is apt to be unpopular. The concept of "sacrifice" is a perceptual one, however, and people often make great sacrifices in order to attain goals that they believe are vital: For instance, think of the privations that people willingly undergo in human warfare, religious movements, or voluntary immigration to a new country. I believe we need to harness this overarching motivation in our efforts to preserve the environment (see next section).

Possible Motivational Approaches

So what positive motivations can we appeal to? Many possibilities are discussed in the following articles in this issue. To begin the dialogue, here are six motivational approaches that cover a wide gamut and may be differentially effective with various types of people.

1. *Voluntary simplicity* as an overall, committed lifestyle has been advocated by Elgin (1993). This means "living lightly on the Earth": a way of life that is outwardly simple and uses the minimum necessary amount of natural resources and technology. In global perspective, it requires all nations to share the Earth's resources efficiently, peacefully, and equitably (Elgin, 1993, p. 42). According to its proponents, this way of life is also inwardly rich: alive, immediate, and poignant in its appreciation of simple experiences and pleasures rather than unneeded material luxuries (pp. 145–146). "It is to live with balance—taking no more than we require and, at the same time, giving fully of ourselves" (p. 157). Currently, it seems that only a few people are willing to make this broad a commitment in all areas of their lives, but the environmental problems facing the world demand that everyone move in that direction. Among the major changes that are necessary for a sustainable world are

   • value changes toward harmony with nature
   • emphasizing long-term goals of environmental protection
   • changing many specific behaviors in a proenvironmental direction
   • publicizing and avoiding short-term environmental damage

   De Young (this issue) has advanced a similar motivational concept, stressing that four types of intrinsic satisfactions are important supports for environmentally responsible behavior: satisfaction from striving for behavioral competence, from frugal, thoughtful consumption, and from participation in valuable community activities, but also from enjoying some of our society's conveniences and luxuries. This viewpoint avoids the implications of self-sacrifice that are often present when altruistic motivations for environmentally responsible behavior are emphasized.

2. Encouraging *specific, concrete actions that are effective* in reducing resource use. This is a promising initial approach because everyone can take some actions easily. For instance, people might install energy-efficient lights, recycle bottles and newspapers, buy recycled products, and carpool or use mass transit. In getting people to do these things, we need to do much more than just *inform* them about desirable actions. We also need to *motivate* them to make behavioral changes, for research shows that simply providing information without strengthening motivation has very little effect (G. T. Gardner & Stern, 1996).

   Many useful guides are available to help people find convenient and practical steps toward reducing their resource use (e.g., EarthWorks Group, 1989). In this issue, excellent examples of effective approaches are presented by Bator and Cialdini and by McKenzie-Mohr. In addition, an organization called Renew America has published an extensive collection of "success stories" about effective community participatory activities that have positive environmental impacts (see Kaplan, this issue).

3. A third motivational approach is *providing clear behavioral norms*, one of the most important ways to create motivation. Norms can come from many sources, but in our society laws and regulations are often crucial in establishing norms. In the environmental area, a few examples are

   • For decades, open burning has not been allowed in Los Angeles County, in order to reduce smog.

   • American cars have to pass increasingly stringent smog checks.

   • Federal fuel-efficiency standards for autos were a great factor in reducing petroleum consumption. Unfortunately, Congress scaled back that objective in the 1980s and also allowed pickup trucks and sport utility vehicles (which are much less fuel-efficient) not to be counted in computing the fleet average.

   • The United States and most other nations of the world have agreed to the Montreal Protocol, phasing out the production of CFCs.

   We need more and more-specific environmental regulations and norms to guide our actions. One very good one was a California requirement that ZEVs (zero-emission vehicles—currently these are electric cars and vehicles powered by fuel cells) had to make up 2% of cars sold in California by 1998. Unfortunately, the California Air Resources Board postponed that requirement in 1996, sending exactly the wrong normative signal.

4. A fourth motivational approach is *harnessing beliefs in technological progress*. Technology can be of great help in using natural resources more efficiently, and we should not reject that as a partial goal. Instead we should emphasize the extreme amount of increased efficiency that is necessary for sustainability. For instance, Olson (1995) proposed a "sustainable community" scenario that involved, not just a doubling of technological efficiency as envisioned by the Brundtland Report, but a 50-fold increase by the year 2050. Though it is not clear whether such massive increases are attainable, there is no doubt that setting them forth as a worldwide goal would be a major impetus to innovation and change. In a similar effort, Lovins (1977, 1980) has spent over 20 years advocating major increases in energy efficiency and demonstrating how they can be realistically achieved.

5. A fifth motivational approach is use of carefully *organized group activity*, which can help to build what Bandura (2000) has termed a sense of collective efficacy. Organized activism is frequently necessary in order to help reduce or

prevent environmental damage, because polluters are very often governments or powerful corporations that can ignore individual complaints. Grassroots political organization and advocacy is difficult, but sometimes it can be remarkably effective in changing public policies in the direction of greater environmental justice for socially disadvantaged groups (cf. Bullard & Johnson, this issue). As mentioned earlier, the mass media are very important influences in developing such group pressures, and every available media avenue needs to be enlisted in the campaign to promote sustainability.

6. A sixth motivational approach is emphasizing the achievement of sustainable living patterns as a superordinate goal that all nations and peoples can share (Sherif, Harvey, White, Hood, & Sherif, 1961)—a *war against the common enemy of an uninhabitable Earth.* This kind of a worldwide campaign can use William James's (1911) concept of a "moral equivalent of war" as the motivational force underlying steps toward the goal of sustainable lifestyles. We need nothing less than this level of fervor in order to accomplish the necessary changes in time to forestall environmental disaster.

Questions for Social Scientists

The points raised above lead me to seven questions about strategies that social scientists should employ in their efforts to achieve sustainable living. I will list them here as a stimulus to thought and discussion:

1. What is the best combination of interventions to use and motivations to appeal to in getting various types of people to adopt environmentally responsible patterns of behavior?

2. How can we best enlist the mass media to promote sustainability and at the same time reject their theme of consumerism and overuse of resources?

3. Should we play down the use of appeals to fear about environmental degradation because we can't offer clear, short-term, effective steps that individuals can take to protect themselves?

4. Should we approach the need for behavior changes by first emphasizing easy, small steps (cf. Weick, 1984) rather than major value changes or reversals in lifestyles?

5. Will it take a major crisis, such as hundreds of millions of deaths, before public opinion is mobilized to action on environmental preservation? And will that happen too late to reverse current trends such as global warming and ozone depletion?

6. How can we best motivate people to avoid the tragedy of the commons—that is, to avoid pursuing their own short-range self-interest at the expense of global, long-term environmental degradation?

7. How can we generate the fervor of a "moral equivalent of war" in our campaign to preserve the environment?

Conclusion

To create a sustainable world, we need everybody's participation. The goal must be to reach a level of environmental impact, as quickly as possible, that will be sustainable for our grandchildren and their grandchildren. Each of us can do many things. We can buy fuel-efficient cars, install extra home insulation instead of running our air conditioners or furnaces, recycle thoroughly, use mass transportation, and limit our families to one or two children. We can also support politicians and other leaders who work for these goals, and we can help to establish social norms by praising our family, friends, and neighbors for their steps toward adopting simple lifestyles. All of us are crucial in this effort. Everything that we can do will benefit our grandchildren—and the whole world.

References

Abernethy, V. D. (1993). *Population politics: The choices that shape our future*. New York: Plenum.

Ausubel, J. H. (1996). Can technology spare the earth? *American Scientist, 84,* 166–178.

Bandura, A. (2000). Exercise of human agency through collective efficacy. *Western Psychologist, 14*(1), 17–20.

Bartelmus, P. (1994). *Environment, growth and development: The concepts and strategies of sustainability*. London: Routledge.

Broder, J. M. (1997, December 12). Clinton adamant on third world role in climate accord. *New York Times*, pp. A1, A16.

Brown, L. R. (1995). Nature's limits. In L. R. Brown et al., *State of the world 1995* (pp. 3–20). New York: Norton.

Brown, L. R. (1999). Feeding nine billion. In L. R. Brown et al., *State of the world 1999* (pp. 115–132). New York: Norton.

Brown, L. R., & Flavin, C. (1999). A new economy for a new century. In L. R. Brown et al., *State of the world 1999* (pp. 3–21). New York: Norton.

Bryant, D., Nielsen, D., & Tangley, L. (1997). *The last frontier forests: Ecosystems and economies on the edge*. Washington, DC: World Resources Institute.

Campbell, C. J., & Laherrere, J. H. (1998). The end of cheap oil. *Scientific American, 278*(3), 78–83.

Clinebell, H. (1998). *Understanding and counseling persons with alcohol, drug, and behavioral addictions*. Nashville, TN: Abingdon.

Cobb, C. W., & Cobb, J. B., Jr. (1994). *The green national product: A proposed index of sustainable economic welfare*. Lanham, MD: University Press of America.

Cohen, J. E. (1995). *How many people can the earth support?* New York: Norton.

Colborn, T., & Clement, C. (Eds.). (1992). *Chemically-induced alterations in sexual and functional development: The wildlife/human connection*. Princeton, NJ: Princeton Scientific.

Colborn, T., Dumanoski, D., & Myers, J. P. (1996). *Our stolen future*. New York: Dutton.

Daly, H. E. (1996). *Beyond growth: The economics of sustainable development*. Boston: Beacon.

David, H. P. (1994). Reproductive rights and reproductive behavior: Clash or convergence of private values and public policies? *American Psychologist, 49*, 343–349.

Dunlap, R., Gallup, G., & Gallup, A. (1993). Global environmental concern: Results from an international public opinion survey. *Environment, 35*, 7–15, 33–39.

EarthWorks Group. (1989). *50 simple things you can do to save the earth.* Berkeley, CA: Earthworks Press.

Ehrlich, P. R., & Ehrlich, A. H. (1991). *Healing the planet.* New York: Addison-Wesley.

Elgin, D. (1993). *Voluntary simplicity: Toward a way of life that is outwardly simple, inwardly rich* (Rev. ed.). New York: Quill.

Fishel, J. (1998). Heated debate in Kyoto. *ZPG Reporter, 30*(1), 4.

Flavin, C., & Dunn, S. (1999). Reinventing the energy system. In L. R. Brown et al., *State of the world 1999* (pp. 22–40). New York: Norton.

French, H. F. (1990). *Clearing the air: A global agenda* (Worldwatch Paper 94). Washington, DC: Worldwatch Institute.

French, H. F. (1997). Learning from the ozone experience. In L. R. Brown et al., *State of the world 1997* (pp. 151–171). New York: Norton.

Gardner, G., & Sampat, P. (1999). Forging a sustainable materials economy. In L. R. Brown et al., *State of the world 1999* (pp. 41–59). New York: Norton.

Gardner, G. T., & Stern, P. C. (1996). *Environmental problems and human behavior.* Boston: Allyn & Bacon.

George, S. (1993). One-third in, two-thirds out. *New Perspectives Quarterly, 10*, 53–55.

Hardin, G. (1968). The tragedy of the commons. *Science, 162*, 1243–1248.

Hardin, G. (1993). *Living within limits: Ecology, economics, and population taboos.* New York: Oxford University Press.

Hileman, B. (1999, August 9). Case grows for climate change. *Chemical & Engineering News*, pp. 16–23.

Intergovernmental Panel on Climate Change (IPCC). (1996). *Climate change 1995: The science of climate change.* New York: Cambridge University Press.

James, W. (1911). The moral equivalent of war. In W. James, *Memories and studies* (pp. 265–296). New York: Longmans, Green.

Katzev, R. D., & Johnson, T. R. (1987). *Promoting energy conservation: An analysis of behavioral research.* Boulder, CO: Westview.

Kempton, W., Boster, J., & Hartley, J. (1995). *Environmental values in American culture.* Cambridge, MA: MIT Press.

Kluckhohn, F. R., & Strodtbeck, F. L. (1961). *Variations in value orientations.* Evanston, IL: Row, Peterson.

Lashof, D. A., & Tirpak, D. A. (1989, February). *Policy options for stabilizing global climate* (Draft report to Congress, Vols. 1 and 2). Washington, DC: U.S. Environmental Protection Agency.

Lemonick, M. D. (1997, November 3). Courting disaster. *Time*, pp. 64–68.

Leventhal, H., Meyer, D., & Nerenz, D. (1980). The common sense representation of illness danger. In S. Rachman (Ed.), *Medical psychology* (Vol. 2, pp. 7–30). New York: Pergamon.

Lovins, A. B. (1977). *Soft energy paths: Toward a durable peace.* London: Pelican.

Lovins, A. B. (1980). Soft energy paths: How to enjoy the inevitable. In *The great ideas today, 1980.* Chicago: Encyclopedia Britannica.

Meadows, D. H., Meadows, D. L., & Randers, J. (1992). *Beyond the limits.* Post Mills, VT: Chelsea Green.

Misch, A. (1994). Assessing environmental health risks. In L. R. Brown et al., *State of the world 1994* (pp. 117–136). New York: Norton.

National Acid Precipitation Assessment Program. (1991). *1990 integrated assessment report.* Washington, DC: Author.

Olson, R. L. (1995). Sustainability as a social vision. *Journal of Social Issues, 51*(4), 15–35.

Oskamp, S. (1995a). Applying social psychology to avoid ecological disaster. *Journal of Social Issues, 51*(4), 217–238.

Oskamp, S. (1995b). Resource conservation and recycling: Behavior and policy. *Journal of Social Issues, 51*(4), 157–177.

Oskamp, S. (2000). A sustainable future for humanity? How can psychology help? *American Psychologist*, *55*, 496–508.

Pimentel, D., Harman, R., Pacenza, M., Pacarsky, J., & Pimentel, M. (1994). Natural resources and an optimum human population. *Population and Environment*, *15*, 347–369.

Schecter, A. (Ed.). (1994). *Dioxins and health*. New York: Plenum.

Schneider, D. (1997, March). The rising seas. *Scientific American*, *276*, 112–117.

Severy, L. J. (Ed.). (1993). *Advances in population: Psychosocial perspectives* (Vol. 1). London: Kingsley.

Sherif, M., Harvey, O. J., White, B. J., Hood, W. R., & Sherif, C. W. (1961). *Intergroup conflict and cooperation: The Robbers Cave experiment*. Norman, OK: University of Oklahoma Book Exchange.

Shireman, W. K. (1993). Solid waste: To recycle or to bury California? In T. Palmer (Ed.), *California's threatened environment: Restoring the dream* (pp. 170–181). Washington, DC: Island Press.

Simon, J. L. (Ed.). (1995). *The state of humanity*. Cambridge, MA: Blackwell.

Stern, P. C., Young, O. R., & Druckman, D. (Eds.). (1992). *Global environmental change: Understanding the human dimensions*. Washington, DC: National Academy Press.

United Nations Development Programme. (1998). *Human development report 1998*. New York: Oxford University Press.

U.S. Environmental Protection Agency. (1992, August). *The consumer's handbook for reducing solid waste* (EPA 530-K-92-003). Washington, DC: Author.

Wackernagel, M., & Rees, W. (1996). *Our ecological footprint: Reducing human impact on the earth*. Philadelphia: New Society.

Weick, K. E. (1984). Small wins: Redefining the scale of social problems. *American Psychologist*, *39*, 40–49.

World Commission on Environment and Development (WCED). (1987). *Our common future*. Oxford, England: Oxford University Press.

World Health Organization. (1992). *Urban air pollution in megacities of the world*. Cambridge, MA: Blackwell.

Wright, L. (1996, January 15). Silent sperm. *New Yorker*, pp. 42–55.

STUART OSKAMP earned his Ph.D. at Stanford and is Professor of Psychology at Claremont Graduate University. His main research interests are in the areas of attitudes and attitude change, environmentally responsible behavior such as recycling and energy conservation, intergroup relations, and social issues and public policy. His books include *Attitudes and Opinions* (1991), *Applied Social Psychology* (1998), and *Social Psychology: An Applied Perspective* (2000). He has been elected president of the American Psychological Association Division of Population and Environmental Psychology and the Society for the Psychological Study of Social Issues, and he has served as editor of the *Journal of Social Issues* and the *Applied Social Psychology Annual* and founding editor of the *Claremont Symposium* series of volumes.

Journal of Social Issues, Vol. 56, No. 3, 2000, pp. 391–406

Empathizing With Nature: The Effects of Perspective Taking on Concern for Environmental Issues

P. Wesley Schultz*

California State University, San Marcos

In this article, I propose that concern for environmental problems is fundamentally linked to the degree to which people view themselves as part of the natural environment. Two studies are reported that test aspects of this theory. The first study describes the structure of people's concern for environmental problems. Results from a confirmatory factor analysis showed a clear three-factor structure, which I labeled egoistic, altruistic, and biospheric. A second study examined the effects of a perspective-taking manipulation on egoistic, social-altruistic, and biospheric environmental concerns. Results showed that participants instructed to take the perspective of an animal being harmed by pollution scored significantly higher in biospheric environmental concerns than participants instructed to remain objective.

In more than 30 years of psychological research, a variety of social psychological theories have been applied to explain attitudes about environmental issues and proenvironmental behavior. One source for theories is social psychological research on prosocial behavior. In this article, I draw on recent theoretical research on altruism and empathy to sketch the beginnings of a broad social-cognitive theory for environmental concern. I argue that the types of environmental concerns people develop are associated with the degree to which they view themselves as interconnected with nature. Data from two studies provide evidence that (1) environmental concerns are clustered into three types and (2) taking the

*Portions of this article were presented at the 1998 meeting of the Society for the Psychological Study of Social Issues, Ann Arbor, MI. I would like to acknowledge the work of Paul Rotkoweitz, Lori Tacey, and Christianna Wolf for their help in conducting these studies. Correspondence concerning this article should be addressed to P. Wesley Schultz, Department of Psychology, California State University, San Marcos, CA 92078 [e-mail: psch@csusm.edu].

perspective of animals being harmed by pollution produces significantly higher levels of concern for the welfare of plants and animals than remaining objective.

In a preceding issue of the *Journal of Social Issues,* Stern and Dietz (1994) proposed that attitudes of environmental concern are rooted in a person's value system (see also Stern, Dietz, & Kalof, 1993, or Stern, Dietz, Kalof, & Guagnano, 1995). They argued that people's attitudes about environmental issues are based on the value that they place on themselves, other people, or plants and animals. Each of these clusters of values provides a distinct basis for environmental concern, such that two people could express the same level of general concern (e.g., concern for air pollution) for fundamentally different reasons (e.g., polluted air is dangerous to my health, polluted air is dangerous to the health of children, or polluted air is damaging to forests). They refer to this model as the value-belief-norm (VBN) theory (see Stern, this issue).

Stern and Dietz (1994) termed these three value-based environmental concerns egoistic, social-altruistic, and biospheric. Egoistic concerns are based on a person's valuing himself or herself above other people and above other living things. "Egoistic values predispose people to protect aspects of the environment that affect them personally, or to oppose protection of the environment if the personal costs are perceived as high" (Stern & Dietz, 1994, p. 70). Although egoistic values are often seen as opposing the environmental movement (Clark, 1995; Oskamp, this issue), it is important to point out that in situations where people high in egoism perceive a threat to themselves from environmental damage, they can be expected to be concerned about environmental problems. Social-altruistic values lead to concern for environmental issues when a person judges environmental issues on the basis of costs to or benefits for other people, be they individuals, a neighborhood, a social network, a country, or all humanity. Biospheric environmental concerns are based on a value for all living things.

A large body of research has linked environmental problems to the human tendency to act in one's own interest (e.g., Bamberg, Kuhnel, & Schmidt, 1999; Diekmann & Preisendorfer, 1998; Hardin, 1968, 1977; Kaiser, Ranney, Hartig, & Bowler, 1999). For example, driving a car a few blocks to the store is beneficial for the individual (e.g., it's faster, requires less physical exertion, and is climate controlled) but is detrimental to the collective (contributes to traffic congestion and noise, uses more natural resources) and detrimental to the environment (air pollution). According to this *rational-choice model*, environmental behavior is motivated by the perceived behavioral consequences associated with various actions. As Batson (1994) points out, however, at times, people do act in ways that increase the welfare of some other person or group of people over self. Indeed, we would expect the rational-choice model to explain more variability in behavior for individuals who place a higher value on self (relative to their valuing of others and of nature) than for individuals who place less relative value on self. Based on the VBN theory summarized above, we would expect the rational-choice model to apply more to egoists than to social-altruists or biospherists.

An argument similar to that made by Stern and Dietz (1994) can be found in Batson (1994) and Batson, Batson, et al. (1995), although Batson does not draw connections between his work and proenvironmental attitudes or behaviors. Expanding on his research concerning empathy and altruism (cf. Batson et al., 1988; Batson et al., 1989; Batson et al., 1991), Batson (1994) points out that at times, people choose to act in the interest of others, even when that action comes at a cost to self. Batson argues that prosocial behavior can be motivated by four different factors: egoism, collectivism, altruism, and principlism. Motives are defined as forces aimed at achieving an ultimate goal, and it is individual differences in these ultimate goals that lead to different motives. These ultimate goals are comparable to Stern and Dietz's (1994) value orientations. For Batson (1994), egoism is a self-interest motive: "a motive is egoistic if the ultimate goal is to increase the actor's own welfare" (p. 604). Choosing to drive a car to a nearby store because it is easier is egoistic. (Similarly, choosing not to drive in order to save money is also egoistic.) Collectivism is a motivation with the ultimate goal of increasing the welfare of a group of people or collective. Altruism is motivation with the ultimate goal of increasing the welfare of "one or more individuals other than oneself" (p. 606). For example, choosing not to drive in order to reduce traffic congestion is altruistic. Finally, principlism is motivation with the ultimate goal of upholding some moral principle. Choosing not to drive in order to improve the quality of life for all living things shows principlism.

The present research builds on the theories of both Stern and Dietz (1994) and Batson and his colleagues (Batson, 1994; Batson, Batson, et al., 1995). Both theories suggest that environmental concerns (which may also serve as motives for behavior) may be clustered around common themes. Following Stern and Dietz (1994), I propose that there are sets of valued objects that are directly linked with environmental concerns. These concerns are based on the negative consequences that could result for valued objects, and these valued objects can be classified as self, other people, or other living things. I refer to these concerns as egoistic, altruistic, and biospheric. Note that I am avoiding the "isms" (e.g., biospherism), because this implies a broad worldview rather than specific attitudes of concern.

I do not assume that these concerns are independent from one another. Instead, I propose that objects are valued because of their perceived relation to self and that egoistic, altruistic, and biospheric concerns reflect varying levels of the inclusiveness of an individual's notion of self (Schultz, 2000). That is, the types of concern for environmental problems that an individual holds are fundamentally linked to the degree to which he or she includes other people and nature within his or her cognitive representations of self. Although a variation of this position was suggested in Dunlap and Van Liere's (1978) more sociological New Environmental Paradigm theory, the theoretical linkages between this notion and current research on environmental concern have not been made. Such a conceptualization offers a broad perspective that could potentially integrate some of the existing research on environmental attitudes and behaviors.

I propose that environmental concern is tied to a person's notion of self and the degree to which people define themselves as independent, interdependent with other people, or interdependent with all living things. From this perspective, concern for environmental issues is an extension of the interconnectedness between two people (Bragg, 1996; Weigert, 1997). We can be interconnected with other people, or more generally, we can be interconnected with all living things. Indeed, the nonscientific literature is replete with references to being "in touch with," "connected with," or "at one with" nature (Hertsgaard, 1999; Nabhan & Trimble, 1994), and stories reflecting an individual's relationship with aspects of the natural world are common across many cultures (cf. Elder & Wong, 1994). People who define themselves as relatively independent from other people and from the natural environment are egoists. They do not view themselves as interconnected with other people or with the natural environment, and so for them, concern for environmental issues will be motivated by reward for the self or the avoidance of harmful consequences (i.e., the rational-choice model prevails). In contrast, environmental concern among people who view themselves as interconnected with others will be based on a desire to gain rewards for people (both specific individuals and people in general) or to avoid harmful consequence for other people. Finally, environmental concern among people who define themselves as part of the biosphere will be based on a desire to gain rewards for all living things or to avoid harmful consequences for the biosphere.

I am not suggesting that individuals with biospheric attitudes are more concerned about environmental problems or that people with egoistic attitudes are unconcerned or apathetic. Indeed, both types of concerns may be predictive of attitudes toward a specific issue, but each has a different foundation. It does seem likely, however, that biospheric concerns provide a broader motive for behavior. For example, we would expect egoistic concerns to be positively predictive of attitudes about specific local issues that directly impact self. In contrast, we would predict that biospheric concerns would be positively related to attitudes about global, more abstract environmental issues, as well as to more specific issues. Thus, we would not be surprised to find people with egoistic and biospheric concerns side by side at a local meeting for the zoning of a landfill. Yet we would not expect people with egoistic concerns to attend a protest to reduce global warming (we would expect to see people with biospheric concerns at such an event).

Based on the social-cognitive theoretical framework sketched above, two studies were conducted. The first study was actually a set of studies designed to test the three-factor model of environmental concern. The second study examined the activation of these concerns by producing an empathic response to different valued objects. Research on prosocial motivation has clearly shown that empathy is a strong predictor of helping behavior. Empathy can be defined as "an other-oriented emotional response congruent with the perceived welfare of another individual" (Batson, Batson, et al., 1995). Extending Batson's empathy-altruism theory to the study of environmental issues, it follows that inducing empathy for the natural environment should lead to the activation of biospheric environmental concerns. The most widely used technique for inducing empathy is perspective taking. Perspective taking is the vicarious experience of another; it is an attempt to understand another person by imagining the other's perspective (Batson, Batson, et al., 1995). Research on perspective taking generally supports the view that "instructions to imagine the affective state of a target frequently trigger a process which ends in the offering of help to that target" (Davis, 1996, p. 145).

Study 1

To assess the clusters of environmental concerns, I identified and tested the factor structure of the valued objects about which people express concern.

Item Development

To identify valued objects, open-ended responses from a recent multinational study were coded (Schultz & Zelezny, 1998). Participants were college students from the United States ($n = 345$), Mexico ($n = 187$), Nicaragua ($n = 78$), Peru ($n = 160$), and Spain ($n = 187$). Participants were asked to complete a four-page questionnaire that contained several established measures of environmental attitudes. As the last item in the questionnaire, participants were asked, "What is the environmental problem that concerns you the most and why?" Respondents were provided with three quarters of a page on which to write their response to this question.

Each open-ended response was coded by a bilingual translator. Responses were coded for (1) the environmental problem listed by the respondent, (2) the object that was harmed by the problem, and (3) the "why" aspect of the response —egoistic, altruistic, or biospheric. The coded responses were then sorted into the three categories, and the seven most often-mentioned valued objects were selected from each of the three value-based groups. The items were then modified so that they were simple and generic enough to be answered by most respondents.

These initial 21 items were administered to a new sample of 245 U.S. undergraduates. Participants were asked to rate each item on a scale from 1 to 7. The introduction stated:

People around the world are generally concerned about environmental problems because of the consequences that result from harming nature. However, people differ in the consequences that concern them the most. Please rate the following items from 1 (not important) to 7 (supreme importance) in response to the question: I am concerned about environmental problems because of the consequences for _____ .

Responses to the 21 items were factor-analyzed using a principal components extraction procedure with a direct oblimin oblique rotation. Through a series of exploratory factor analyses, 12 items (four each from egoistic, altruistic, and biospheric) that generated a clear three-factor structure were identified. Selection of these 12 items was based on factor loadings, commonalities, zero-order correlation coefficients, and theoretical grounds.

The 12 items were then factor-analyzed a second time and rotated using a direct oblimin procedure. Three factors with eigenvalues greater than 1.0 were extracted that accounted for 74% of the total variance. Factor loadings for the three extracted factors are presented in Table 1. The first factor represents a biospheric factor, and the items with strong factor loadings were "marine life," "birds," "animals," and "plants." The second factor represents an egoistic factor; the variables with strong factor loadings were "my health," "my future," "my lifestyle," and "me." The third factor was labeled altruistic, and it was defined by "children," "people in my community," "all people," and "my children." Correlations between the three factors were $r = .25$ for egoistic and biospheric, $r = .37$ for biocentric and altruistic, and $r = .39$ for egoistic and altruistic. To further examine my proposed three-factor model, a confirmatory factor analysis (CFA) was performed with a new sample.

Confirmatory Factor Analysis

Sample. Participants in the study were 400 psychology undergraduates from the United States. Participants rated the 12 environmental items identified above.

Statistical analysis. A CFA was performed using AMOS 3.6. Missing values were replaced with series means.

Table 1. Egoistic, Social-Altruistic, and Biospheric Scale Items and Rotated Factor Loadings

Scale and item	Rotated factor loadings[a]		
	Factor 1	Factor 2	Factor 3
Biospheric concerns			
Animals	.22	**.90**	.36
Plants	.27	**.85**	.29
Marine life	.21	**.93**	.35
Birds	.26	**.93**	.44
Egoistic concerns			
Me	**.89**	.18	.44
My future	**.83**	.24	.40
My lifestyle	**.78**	.12	.24
My health	**.80**	.39	.30
Altruistic concerns			
All people	.31	.33	**.75**
Children	.23	.37	**.76**
People in my community	.38	.30	**.90**
My children	.41	.28	**.93**

[a]An oblimin rotation was used. Factor loadings shown are from the rotated matrix.

Results. The CFA tested three possible models: a one-factor, a two-factor, and a three-factor model. The one-factor model is consistent with the view of environmental concern as a unidimensional construct ranging from *unconcerned* at the low end, to *concerned* at the high end. This is the implicit model adopted in much of the research on attitudes of environmental concern. To test the one-factor model, all 12 environmental items were loaded on a single factor. The two-factor model is consistent with the classification of environmental attitudes as rooted either in a concern for all living things or in a concern for humans (self included; cf. Thompson & Barton, 1994). To test this model, the four biospheric concerns were loaded on one factor, and the remaining eight items (four egoistic and four social-altruistic) were loaded on a second factor. The three-factor model is consistent with Stern and Dietz's (1994) tripartite conceptualization of environmental concerns grounded in clusters of valued objects. I expected the three-factor model to provide the best overall fit to the data.

Results are based on maximum likelihood estimates produced from covariance matrices. Analyses indicated that the independence model could be rejected (df = 66, χ^2 = 2200.28, χ^2/df = 33.37, root mean-square error of approximation [RMSEA] = .29, goodness-of-fit index [GFI] = .38, adjusted GFI [AGFI] = .26, Tucker-Lewis index [TLI] = .00). The one-factor model showed an improved, but still unacceptable fit (df = 54, χ^2 = 821.69, χ^2/df = 15.22, RMSEA = .19, GFI = .68, AGFI = .54, TLI = .56). The two-factor model was significantly better, $\chi^2(1)$ = 406.47, $p < .001$, than the one-factor model (df = 53, χ^2 = 415.22, χ^2/df = 7.83, RMSEA = .13, GFI = .83, AGFI = .75, TLI = .80), but did not provide an acceptable fit—all of the fit indices were beyond my established limits. The three-factor model showed a significant, $\chi^2(2)$ = 196.74, $p < .001$, improvement over the two-factor model and provided an overall acceptable fit (df = 51, χ^2 = 218.48, χ^2/df = 4.28, RMSEA = .08, GFI = .92, AGFI = .90, TLI = .90). The unstandardized factor weights, standardized factor weights (shown in parentheses), covariances between the three factors, and correlation coefficients between the three factors (shown in parentheses) are presented in Figure 1.

Study 2

The results from Study 1 showed support for the distinction between egoistic, altruistic, and biospheric attitudes of environmental concern. The second study was an experimental attempt to activate different environmental concerns using a perspective-taking manipulation. I have argued that the types of concerns an individual has for environmental problems are associated with the degree to which the individual includes nature within his or her cognitive representations of self. Based on this perspective, I predicted that taking the perspective of another person or an animal would lead to a greater inclusiveness and subsequently, greater levels of biospheric environmental concern.

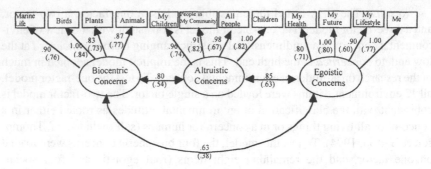

Fig. 1. Results from a confirmatory factor analysis of the three-factor structure

Methodology

Sample. Participants in the study were 180 undergraduates recruited from the psychology department's human participant pool.

Materials. Participants were randomly assigned to view one of three sets of pictures: people engaging in recreational activities in a natural environment (a woman meditating on the beach, a hiker in a forest, a painter near a lake, a rock climber, a person canoeing), animals in a natural environment (a caribou on a hill, gorillas in a forest, a bear, a rhinoceros on a savanna, a breaching whale), or animals being harmed by nature (a seal caught in a fishing net, an eagle on a smoky factory smokestack, an otter in an oil spill, a bear in a trash pile, a bird with a plastic bag around its neck). Color images were shown on a 15" SVGA color monitor in 8-bit color using Microsoft PowerPoint 4.0. Each image appeared on the screen for 30 s; participants viewed a total of five slides.

After viewing the slides, participants completed a questionnaire that contained several measures of environmental attitudes, including the 12 environmental concern items identified above. Separate scale scores were produced for egoistic, altruistic, and biospheric concerns by averaging the four items in each domain. Alpha reliabilities for the three subscales were all high: egoistic (alpha = .91), altruistic (alpha = .92), and biospheric (alpha = .94).

Procedure. The study was conducted in a small laboratory room. Upon arrival, participants provided informed consent and were given written instructions that were read aloud by the experimenter. Half of the participants were randomly assigned to an "objective" condition. Instructions read:

> As you view the images, look closely at the subjects within each image. Make careful observations about the subjects' mannerisms, postures, movements, and facial expressions. Notice exactly what the subject is doing, whatever it is. Try to take a neutral perspective, being as objective as possible about the subjects. Do not concern yourself with feelings or

views. Do not let yourself become caught up in imagining what the subject has been through. Just concentrate on the images objectively.

The other half of the participants were assigned to a "perspective-taking" condition. Instructions read:

> As you view the images, try to imagine how the subjects in the images feel. Try to take the perspective of the subjects, imagining how they are feeling about what is happening. While you view the images, picture to yourself just how they feel. Think about their reactions. In your mind's eye visualize clearly and vividly how they feel in their situation. Try not to concern yourself with attending to all the information presented. Just imagine how the subjects feel in their situation.

After participants indicated that they understood the instructions, the lights were dimmed and the slide show began. The first slide informed the participants about the types of slides they would see.

As part of the postsession questionnaire, participants completed four manipulation check items: To what extent did you try to imagine how the subjects were feeling? To what extent did you objectively observe the subjects in the images? To what extent did you take the perspective of the subjects in the images? To what extent did you remain detached from the subjects in the images? Items were rated on a 5-point scale from 1 (*not at all*) to 5 (*all of the time*).

Results

This was a 2×3 factorial experiment with 30 participants in each cell. Preliminary analyses examined the scores on the four manipulation check items. Four 2×3 analyses of variance (ANOVAs) were performed: one for each manipulation check item. Across all four analyses, the results showed a significant main effect for perspective taking, no main effect for picture type, and no interaction. Univariate tests revealed significant differences in the expected direction for three of the four manipulation check items (eta-squared = .54, .36, and .21 for items 1, 3, and 4, respectively). For the second item, "To what extent did you objectively observe the subjects in the images?" the results showed a nonsignificant difference between the perspective-taking ($M = 4.01$) and the objective ($M = 4.02$) condition, $F(1, 174) = .01, p = 93$, eta-squared = .00. I attribute this to awkward wording of the item; participants in the perspective condition may have interpreted "objectively observe" to mean "look carefully" and subsequently indicated that they did.

Responses to the 12 environmental-concern items were analyzed using a 2 (perspective, objective) $\times$ 3 (picture type) multivariate analysis of variance (MANOVA), with egoistic, altruistic, and biospheric concerns as the dependent variables. The results revealed a significant multivariate picture type by perspective-taking interaction, $F(6, 346) = 2.77, p = .01$, Pillais = .09. Neither the main effect for picture type, $F(6, 346) = 1.65, p = .13$, nor the main effect of perspective taking, $F(3, 172) = .53, p = .66$, was significant. Follow-up 2×3 univariate tests for each of the three dependent variables revealed a significant interaction for biospheric ($F = 4.44, p = .013$) and for altruistic ($F = 5.95, p = .003$) concerns, but not for egoistic concerns ($F = 1.33, p = .27$).

For biospheric concerns, the interaction showed that when the picture was an animal being harmed by pollution, participants in the perspective-taking condition scored significantly higher ($M = 5.82$) than participants in the objective condition ($M = 5.01$), $F(1, 178) = 5.34, p = .02$. No significant differences were observed between the perspective-taking and objective conditions when the image was an animal in nature, $F(1, 178) = .01$, ns. A marginally significant difference was observed when the image was a person in nature, $F(1, 178) = 3.54, p = .06$, with the perspective-taking condition scoring lower ($M = 4.93$) than the objective condition ($M = 5.59$). The mean biospheric concern scores are shown in Figure 2.

For altruistic concerns, the interaction revealed that for pictures of animals being harmed, perspective taking produced significantly higher scores ($M = 6.01$) than remaining objective ($M = 5.19$), $F(1, 178) = 5.02, p = .03$. No significant differences were found, however, between the perspective-taking and objective conditions for either the animals in nature or the people in nature conditions.

Discussion

As the problems associated with pollution, overpopulation, energy consumption, overuse of natural resources, and other environmental issues become more

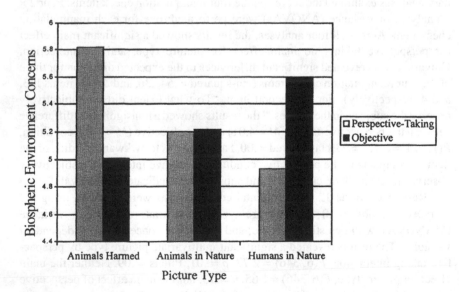

Fig. 2. Mean biospherism scores as a function of perspective taking and picture type

pressing, one might think that psychologists would step forward with models and theories for understanding environmental concerns, motives, and behaviors. Unfortunately, we have not. Although the psychological study of environmental issues has produced some interesting and useful findings, the bulk of the research tends to be fragmented and difficult to integrate into an organized theory. Much of the research on environmental issues has been based on traditional social psychological theories of attitudes.

One promising theoretical approach to the study of environmental attitudes is the value-basis theory. As articulated by Stern and Dietz (1994) and Stern et al. (1995), the value-basis theory for environmental concern proposes that attitudes are formed by considering a few salient aspects of an attitude object and the relevance of this object to a few salient values. Research in this area has been based on Schwartz's (1977) theory for normative decision making and has focused on an altruistic value. Environmental concerns and behaviors are viewed as the result of an activated altruistic moral norm (e.g., Black, Stern, & Elworth, 1985; Hopper & Nielson, 1991; Schultz & Zelezny, 1998). Building on this work, the value-basis theory proposes that attitudes toward environmental issues can be linked to a broader range of values, not just altruism. Concern for environmental issues can be based on the relevance of environmental damage to self, people, or all living things. The type of concern that develops depends largely on the relevance of attitude objects to activated values.

My approach in the studies described in this article is consistent with the value-basis theory, but my focus was on clusters of valued objects rather than on values per se. That is, I measured specific attitude objects (e.g., concern for plants, people, self) and not values (e.g., equality, loyalty, broad-mindedness, a world of beauty). (See Karp, 1996; Schultz and Zelezny, 1999; or Stern et al., 1995, for data on the relationship between values and environmental attitudes.) The findings reported in this article showed that there are distinct clusters of environmental attitudes: biocentric concerns focus on all living things (plants, marine life, birds, animals), altruistic concerns focus on other people (people in my community, children, all people, my children), and egoistic concerns focus on the self (my health, my future, my lifestyle, me, and my prosperity). Among a sample of college students, we found evidence for this three-factor model. A similar three-factor structure has been found in an international sample of college students and a sample of California residents (Schultz, 2000).

These results are consistent with Stern and Dietz's (1994) value-basis theory. I further propose, however, not only that these concerns are organized around valued objects, but that these objects are valued because they are included in a person's cognitive representation of self. In 1978, Dunlap and Van Liere proposed that a New Environmental Paradigm (NEP) was emerging in which people viewed humans as an integral part of nature. Subsequent research proceeded to examine correlates of NEP, particularly attitudes and behaviors. The NEP, however, measures an individual's perception of the relationship between humans and the natural environment. Stern et al. (1995) have suggested that the NEP reflects a general awareness of the consequences of harming nature. The NEP, in its focus on "humans," is more sociological than psychological.

In this article, I have sketched a psychological variation on the NEP: a social-cognitive perspective that is consistent with several current areas of social psychological research (Aron, Aron, Tudor, & Nelson, 1991; Aron & Fraley, 1999; Batson, 1994; Stern & Dietz, 1994). I have argued that environmental concerns are rooted in a person's interconnection with other people and with the natural environment. I do not suggest that this is a disposition. Indeed, just as a relationship between two people can deepen and become more "interconnected," so too can our relationship with the natural environment.

The results from Study 2 provide some evidence that environmental concerns are malleable across situations. The results showed that when viewing images of animals being harmed by nature, participants instructed to take the animals' perspective expressed significantly higher levels of biospheric environmental concerns than participants instructed to remain objective. To my knowledge, this is the first reported laboratory experiment in which environmental attitudes have been used as a dependent variable; most studies have used environmental concern as a predictor of other attitudes or behaviors or as the criterion variable in studies not involving an experimental manipulation (e.g., as predicted by political ideology, gender, income, and so on). A few notable exceptions can be found in the environmental education literature in which environmental concerns are measured following an educational activity (for reviews of this literature, see Dwyer, Leeming, Cobern, Porter, & Jackson, 1993; Leeming, Dwyer, Porter, & Cobern, 1993; or Zelezny, 1999).

At this point, it might not be clear why taking the perspective of an animal being harmed by nature would produce an increase in biospheric concerns. One potential explanation for this finding comes from research on perspective taking and empathy (Batson, Turk, Shaw, & Klein, 1995; Davis, Conklin, Smith, & Luce, 1996; Dovidio, Allen, & Schroeder, 1990; Eisenberg & Miller, 1987). The empathy-altruism hypothesis predicts that helping behavior can be produced by both egoistic and altruistic motives. Taking the perspective of a person being harmed leads to empathy and to the activation of an altruistic motive. Empathy is defined as other-oriented feelings of concern about the perceived welfare of another person. In contrast, if the other's perspective is not taken, then empathy is not induced, and the egoistic motive is dominant. Both motives can lead to helping behavior: altruism for no obvious benefit for self, and egoism to gain reward or to avoid punishment for self. This line of reasoning suggests that our perspective-taking manipulation may have generated feelings of empathy and subsequently a greater concern for the welfare of animals and the biosphere.

In addition to producing feelings of empathy, taking perspective may also have temporarily increased the extent to which participants viewed themselves as interconnected with nature. That is, taking perspective may have expanded the participants' inclusiveness of self and reduced the degree of separation that participants perceived between themselves and nature. There is some evidence that a perspective-taking manipulation can have such an effect. Davis et al. (1996) demonstrated that taking the perspective of another person produced a greater degree of other-inclusion in self. That is, when we take the perspective of another person, we expand our boundaries of self to include the other. Davis et al. (1996) showed that experimentally manipulating perspective taking caused observers to create cognitive views of other that overlapped with the observer's own self-representations. Experimentally manipulating perspective taking produced a greater inclusion of other in self (see also Aron et al., 1991, and Aron & Fraley, 1999, for an examination of the changes in the degree of inclusion of other in self associated with intimate interpersonal relationships). Building on this finding, I view perspective taking as a manipulation of the interconnectedness between self, other, and biosphere.

This perspective on environmental concerns leads to some additional hypotheses for future research. First, attitudes of environmental concern should be positively correlated with measures of empathy, especially empathy scores that focus on perspective taking. Second, it should be possible to assess the content of self-schemata and identify differences in the degree to which people include nature in their cognitive representation of self. Third, it should be possible to experimentally manipulate threats to valued objects (self, other people, plants, and animals) and show predictable patterns of helping behavior for people high in egoistic, altruistic, or biospheric concerns. Finally, it should be possible to develop interventions (cf. Bator & Cialdini, this issue; McKenzie-Mohr, this issue) or environmental education programs that evoke feelings of empathy or inclusion and lead to biospheric environmental concerns.

This approach is especially applicable to environmental education activities. My results suggest that any activity that reduces an individual's perceived separation between self and nature will lead to an increase in that individual's biospheric concern. For example, a hike in the woods, a class trip to a natural park, a family camping trip (in a tent, not a recreational vehicle), an animal presentation in which students can see and touch the animal, or creating birdhouses or gardens should all lead to greater interconnectedness and inclusion. By contrast, a trip to a zoo to see animals in cages, watching animals perform skits or trained shows, hearing information about animals or nature taught abstractly in a classroom, or environmentally destructive recreational behaviors (like off-road motorcycles, jet skis, and snowmobiles) will likely lead to less perceived interconnection and more egoistic attitudes about nature.

Conclusion

In this article, I have demonstrated the existence of a clear three-factor structure for environmental concerns that I labeled egoistic, altruistic, and biospheric. These findings are consistent with Stern's value-basis theory for environmental attitudes. As an extension, I have attempted to link environmental concern to the inclusion of others in self and to the inclusion of nature in self. Further, I have proposed that these concerns are associated with empathy and that greater levels of inclusion can be produced by taking the perspective of animals being harmed by nature (biospheric) or people being harmed by nature (altruistic). I believe that this conceptualization offers a promising new avenue for basic research on environmental concern and also a useful theory for applied research on encouraging proenvironmental behavior.

References

Aron, A., Aron, E. N., & Smollan, D. (1992). Inclusion of other in the self scale and the structure of interpersonal closeness. *Journal of Personality and Social Psychology, 63,* 596–612.

Aron, A., Aron, E. N., Tudor, M., & Nelson, G. (1991). Close relationships as including other in the self. *Journal of Personality and Social Psychology, 60,* 241–253.

Aron, A., & Fraley, B. (1999). Relationship closeness as including other in the self: Cognitive underpinnings and measures. *Social Cognition, 17,* 140–160.

Bamberg, S., Kuhnel, S. M., & Schmidt, P. (1999). The impact of general attitude on decisions: A framing approach. *Rationality and Society, 11,* 5–25.

Batson, C. D. (1991). *The altruism question.* Hillsdale, NJ: Erlbaum.

Batson, C. D. (1994). Why act for the public good? Four answers. *Personality and Social Psychology Bulletin, 20,* 603–610.

Batson, C. D., Batson, J. G., Griffitt, C. A., Barrientos, S., Brandt, J. R., Sprengelmeyer, P., & Bayly, M. J. (1989). Negative-state relief and the empathy-altruism hypothesis. *Journal of Personality and Social Psychology, 56,* 922–933.

Batson, C. D., Batson, J. G., Slingsby, J. K., Harrell, K. L., Peekna, H. M., & Todd, R. M. (1991). Empathic joy and the empathy-altruism hypothesis. *Journal of Personality and Social Psychology, 61,* 413–426.

Batson, C. D., Batson, J., Todd, R. M., Brummett, B. H., Shaw, L. L., & Aldeguer, C. M. R. (1995). Empathy and the collective good: Caring for one of the others in a social dilemma. *Journal of Personality and Social Psychology, 68,* 619–631.

Batson, C. D., Dyck, J. L., Brandt, J. R., Batson, J. G., Powell, A. L., McMaster, M. R., & Griffitt, C. (1988). Five studies testing two new egoistic alternatives to the empathy-altruism hypothesis. *Journal of Personality and Social Psychology, 55,* 52–77.

Batson, C. D., Turk, C. L., Shaw, L. L., & Klein, T. R. (1995). Information function of empathic emotion: Learning that we value the other's welfare. *Journal of Personality and Social Psychology, 68,* 300–313.

Black, J. S., Stern, P. C., & Elworth, J. T. (1985). Personal and contextual influences on household energy adaptations. *Journal of Applied Psychology, 70,* 3–21.

Bragg, E. A. (1996). Towards ecological self: Deep ecology meets constructionist self-theory. *Journal of Environmental Psychology, 16,* 93–108.

Clark, M. E. (1995). Changes in Euro-American values needed for sustainability. *Journal of Social Issues, 51*(4), 63–82.

Davis, M. H. (1996). *Empathy: A social psychological approach.* New York: Westview.

Davis, M. H., Conklin, L., Smith, A., & Luce, C. (1996). Effect of perspective taking on the cognitive representation of persons: A merging of self and other. *Journal of Personality and Social Psychology, 70,* 713–726.

Diekmann, A., & Preisendorfer, P. (1998). Environmental behavior: Discrepancies between aspirations and reality. *Rationality and Society, 10,* 79–103.

Dietz, T., & Stern, P. C. (1995). Toward a theory of choice: Socially embedded preference construction. *Journal of Socio-Economics, 24,* 261–279.

Dovidio, J. F., Allen, J. L., & Schroeder, D. A. (1990). Specificity of empathy-induced helping: Evidence for altruistic motivation. *Journal of Personality and Social Psychology, 59,* 249–260.

Dunlap, R., & Van Liere, K. (1978). The new environmental paradigm. *Journal of Environmental Education, 9*(4), 10–19.

Dunlap, R., Van Liere, K., Mertig, A., & Howell, R. (1992, August). *Measuring endorsement of an ecological worldview: A revised NEP scale.* Paper presented at the Annual Meeting of the Rural Sociology Society, State College, PA.

Dwyer, W. O., Leeming, F. C., Cobern, M. K., Porter, B. E., & Jackson, J. N. (1993). Critical review of behavioral interventions to preserve the environment: Research since 1980. *Environment and Behavior, 25,* 275–321.

Eisenberg, N., & Miller, P. A. (1987). Empathy and prosocial behavior. *Psychological Bulletin, 101,* 91–119.

Elder, J., & Wong, H. D. (Eds.). (1994). *Family of earth and sky: Indigenous tales of nature from around the world.* Boston: Beacon.

Guagnano, G., Dietz, T., & Stern, P. C. (1994). Willingness to pay for public goods: A test of the contribution model. *Psychological Science, 5,* 411–415.

Hardin, G. (1968). The tragedy of the commons. *Science, 162,* 1243–1248.

Hardin, G. (1977). *The limits of altruism: An ecologist's view of survival.* Bloomington, IN: Indiana University Press.

Hertsgaard, M. (1999). *Earth odyssey: Around the world in search of our environmental future.* New York: Broadway Books.

Hopper, J., & Nielson, J. M. (1991). Recycling as altruistic behavior: Normative and behavioral strategies to expand participation in a community curbside recycling program. *Environment and Behavior, 23,* 195–220.

Kaiser, F., Ranney, M., Hartig, T., & Bowler, P. A. (1999). Ecological behavior, environmental attitude, and feelings of responsibility for the environment. *Journal of Environmental Psychology, 4,* 59–74.

Karp, D. G. (1996). Values and their effect on pro-environmental behaviour. *Environment and Behavior, 28,* 111–133.

Leeming, F. C., Dwyer, W. O., Porter, B. E., & Cobern, M. K. (1993). Outcome research in environmental education: A critical review. *Journal of Environmental Education, 24*(4), 8–21.

Merchant, C. (1992). *Radical ecology: The search for a livable world.* New York: Routledge.

Nabhan, G. P., & Trimble, S. (1994). *The geography of childhood: Why children need wild places.* Boston: Beacon.

Schultz, P. W. (2000). Assessing the structure of environmental concern: Concern for self, other people, and the biosphere. Unpublished manuscript.

Schultz, P. W., & Zelezny, L. C. (1996, June). *Culture, values, and concern for the environment: A cross-cultural study.* Paper presented at the Society for the Psychological Study of Social Issues, Ann Arbor, MI.

Schultz, P. W., & Zelezny, L. C. (1998). Values and proenvironmental behavior: A five-country survey. *Journal of Cross-Cultural Psychology, 29,* 540–558.

Schultz, P. W., & Zelezny, L. C. (1999). Values as predictors of environmental attitudes: Evidence for consistency across cultures. *Journal of Environmental Psychology, 19,* 255–265.

Schwartz, S. H. (1977). Normative influences on altruism. In L. Berkowitz (Ed.), *Advances in experimental social psychology* (Vol. 10, pp. 221–279). New York: Academic.

Stern, P. C., & Dietz, T. (1994). The value basis of environmental concern. *Journal of Social Issues, 50,* 65–84.

Stern, P. C., Dietz, T., & Kalof, L. (1993). Value orientations, gender, and environmental concern. *Environment and Behavior, 25,* 322–348.

Stern, P. C., Dietz, T., Kalof, L., & Guagnano, G. A. (1995). Values, beliefs, and proenvironmental action: Attitude formation toward emergent attitude objects. *Journal of Applied Social Psychology, 25,* 1611–1636.

Thompson, S. C. G., & Barton, M. A. (1994). Ecocentric and anthropocentric attitudes toward the environment. *Journal of Environmental Psychology, 14,* 149–157.

Weigert, A. J. (1997). *Self, interaction, and natural environment: Refocusing our eyesight.* New York: State University of New York Press.

Zelezny, L. C. (1999). Educational interventions that improve environmental behaviors: A meta-analysis. *Journal of Environmental Education, 31,* 5–14.

P. WESLEY SCHULTZ is an Associate Professor of Psychology at California State University, San Marcos, where he teaches courses in psychology and statistics. His research interests are in the psychology of environmental issues and the application of psychological theory to help understand and solve social problems. Recent projects have focused on promoting recycling, cross-cultural research on environmental attitudes, environmental education, stereotypes, and prejudice. He is coauthor of *Applied Social Psychology* (1998, Prentice-Hall) and *Social Psychology: An Applied Perspective* (2000, Prentice-Hall) and numerous research publications in the areas of social psychology and environmental issues.

Journal of Social Issues, Vol. 56, No. 3, 2000, pp. 407–424

Toward a Coherent Theory of Environmentally Significant Behavior

Paul C. Stern

National Research Council

This article develops a conceptual framework for advancing theories of environmentally significant individual behavior and reports on the attempts of the author's research group and others to develop such a theory. It discusses definitions of environmentally significant behavior; classifies the behaviors and their causes; assesses theories of environmentalism, focusing especially on value-belief-norm theory; evaluates the relationship between environmental concern and behavior; and summarizes evidence on the factors that determine environmentally significant behaviors and that can effectively alter them. The article concludes by presenting some major propositions supported by available research and some principles for guiding future research and informing the design of behavioral programs for environmental protection.

Recent developments in theory and research give hope for building the understanding needed to effectively alter human behaviors that contribute to environmental problems. This article develops a conceptual framework for the theory of environmentally significant individual behavior, reports on developments toward such a theory, and addresses five issues critical to building a theory that can inform efforts to promote proenvironmental behavior.

---

*This research was supported in part by the U.S. Environmental Protection Agency grant, "The Social Psychology of Stated Preferences," and by National Science Foundation grants SES 9211591 and 9224036 to George Mason University. I thank my colleagues Gregory Guagnano, Linda Kalof, and especially Thomas Dietz and Gerald Gardner for their collaboration, support, and criticism in our collective effort to theorize about environmental concern and behavior. Correspondence concerning this article should be addressed to Paul C. Stern, National Research Council, 2101 Constitution Ave., N.W. (HA-172), Washington DC 20418 [e-mail: pstern@nas.edu].

Defining Environmentally Significant Behavior

Environmentally significant behavior can reasonably be defined by its impact: the extent to which it changes the availability of materials or energy from the environment or alters the structure and dynamics of ecosystems or the biosphere itself (see Stern, 1997). Some behavior, such as clearing forest or disposing of household waste, directly or proximally causes environmental change (Stern, Young, & Druckman, 1992). Other behavior is environmentally significant indirectly, by shaping the context in which choices are made that directly cause environmental change (e.g., Rosa & Dietz, 1998; Vayda, 1988). For example, behaviors that affect international development policies, commodity prices on world markets, and national environmental and tax policies can have greater environmental impact indirectly than behaviors that directly change the environment.

Through human history, environmental impact has largely been a by-product of human desires for physical comfort, mobility, relief from labor, enjoyment, power, status, personal security, maintenance of tradition and family, and so forth, and of the organizations and technologies humanity has created to meet these desires. Only relatively recently has environmental protection become an important consideration in human decision making. This development has given environmentally significant behavior a second meaning. It can now be defined from the actor's standpoint as behavior that is undertaken with the intention to change (normally, to benefit) the environment. This intent-oriented definition is not the same as the impact-oriented one in two important ways: It highlights environmental intent as an independent cause of behavior, and it highlights the possibility that environmental intent may fail to result in environmental impact. For example, many people in the United States believe that avoiding the use of spray cans protects the ozone layer, even though ozone-destroying substances have been banned from spray cans for two decades. The possible discrepancy between environmental intent and environmental impact raises important research questions about the nature and determinants of people's beliefs about the environmental significance of behaviors.

Both definitions of environmentally significant behavior are important for research but for different purposes. It is necessary to adopt an impact-oriented definition to identify and target behaviors that can make a large difference to the environment (Stern & Gardner, 1981a). This focus is critical for making research useful. It is necessary to adopt an intent-oriented definition that focuses on people's beliefs, motives, and so forth in order to understand and change the target behaviors.

Types of Environmentally Significant Behavior

Much early research on proenvironmental behavior presumed it to be a unitary, undifferentiated class. More recently it has become clear that there are several distinct types of environmentally significant behavior and that different combinations of causal factors determine the different types.

Environmental Activism

Committed environmental activism (e.g., active involvement in environmental organizations and demonstrations) is a major focus of research on social movement participation. This research provides detailed analysis of the "recruitment" process through which individuals become activists (McAdam, McCarthy, & Zald, 1988).

Nonactivist Behaviors in the Public Sphere

Recently, the social movement literature has pointed to nonactivists' support of movement objectives as another important class of behavior (Zald, 1992). Public opinion researchers and political scientists sometimes examine such behavior, but relatively little research has been done to classify the behaviors into coherent subtypes. It seems reasonable as a first approximation to distinguish between more active kinds of environmental citizenship (e.g., petitioning on environmental issues, joining and contributing to environmental organizations) and support or acceptance of public policies (e.g., stated approval of environmental regulations, willingness to pay higher taxes for environmental protection). My colleagues and I have found empirical support for distinguishing these types from each other and from activism (Dietz, Stern, & Guagnano, 1998; Stern, Dietz, Abel, Guagnano, & Kalof, 1999). Although these behaviors affect the environment only indirectly, by influencing public policies, the effects may be large, because public policies can change the behaviors of many people and organizations at once. An important feature of public-sphere behaviors, including activism, is that environmental concerns are within awareness and may therefore be influential.

Private-Sphere Environmentalism

Consumer researchers and psychologists have focused mainly on behaviors in the private sphere: the purchase, use, and disposal of personal and household products that have environmental impact. It is useful to subdivide these according to the type of decision they involve: the purchase of major household goods and services that are environmentally significant in their impact (e.g., automobiles, energy for the home, recreational travel), the use and maintenance of environmentally important goods (e.g., home heating and cooling systems), household waste disposal, and "green" consumerism (purchasing practices that consider the environmental impact of production processes, for example, purchasing recycled products and organically grown foods). Making such distinctions has revealed that some types of choice, such as infrequent decisions to purchase automobiles and major household appliances, tend to have much greater environmental impact than others, such as changes in the level of use of the same equipment: the distinction between efficiency and curtailment behaviors (Stern & Gardner, 1981a, 1981b). Private-sphere behaviors may also form coherent clusters empirically (e.g., Bratt, 1999a), and different types of private-sphere behavior may have different determinants (e.g., Black, Stern, & Elworth, 1985). Private-sphere behaviors are unlike public-sphere environmentalism in that they have direct environmental consequences. The environmental impact of any individual's personal behavior, however, is small. Such individual behaviors have environmentally significant impact only in the aggregate, when many people independently do the same things.

Other Environmentally Significant Behaviors

Individuals may significantly affect the environment through other behaviors, such as influencing the actions of organizations to which they belong. For example, engineers may design manufactured products in more or less environmentally benign ways, bankers and developers may use or ignore environmental criteria in their decisions, and maintenance workers' actions may reduce or increase the pollution produced by manufacturing plants or commercial buildings. Such behaviors can have great environmental impact because organizational actions are the largest direct sources of many environmental problems (Stern & Gardner, 1981a, 1981b; Stern, 2000). The determinants of individual behavior within organizations are likely to be different from those of political or household behaviors.

Evidence for Distinguishing Major Behavioral Types

Research my colleagues and I have conducted suggests that this distinction among behavioral types is not only conceptually coherent but statistically reliable and psychologically meaningful. For instance, a factor analysis of the behavioral items in the environment module of the 1993 General Social Survey revealed a three-factor solution (Dietz et al., 1998). One factor included four private-sector household behaviors (e.g., buying organic produce, sorting household waste for recycling); a second included two environmental citizenship behaviors (signing a petition and belonging to an environmental group); and the third included three items indicating willingness to make personal financial sacrifices for environmental goals, which assess policy support. A different pattern of social-psychological and socio-demographic predictors was associated with each of the behavioral types, and even the two citizenship behaviors had quite different sets of predictors.

My colleagues and I had similar results using data from a 1994 national environmental survey (Stern et al., 1999). Factor analysis of 17 items measuring self-reported behaviors and behavioral intentions again revealed three factors: consumer behaviors (e.g., buying organic produce, avoiding purchases from companies that harm the environment); environmental citizenship (e.g., voting, writing to government officials); and policy support, expressed as willingness to sacrifice economically to protect the environment (e.g., by paying much higher taxes or prices). Self-reported participation in environmental demonstrations and protests, presumably a measure of committed activism, did not load on any of the above three factors. Each of these factors was predicted by a different pattern of norms, beliefs, and values, and activism had yet a different set of predictors.

The Determinants of Environmentalism

Environmentalism may be defined behaviorally as the propensity to take actions with proenvironmental intent. Some theories treat environmentalism as a matter of worldview. Perhaps the most prominent example in social psychology is the idea that it flows from adopting a New Environmental (or Ecological) Paradigm, within which human activity and a fragile biosphere are seen as inextricably interconnected (Dunlap, Van Liere, Mertig, & Jones, this issue). Another worldview theory explains environmentalism in terms of an egalitarian "cultural bias" or "orienting disposition" (Dake, 1991; Douglas & Wildavsky, 1982; Steg & Sievers, 2000). Recently, some researchers have begun to explore affective influences on environmental concern and behavior, including sympathy for others (Allen & Ferrand, 1999), "emotional affinity" toward nature (Kals, Schumacher, & Montada, 1999), and empathy with wild animals (Schultz, this issue).

Some theories look to values as the basis of environmentalism. Inglehart (1990) suggests that it is an expression of postmaterialist values of quality of life and self-expression that emerge as a result of increasing affluence and security in the developed countries. Some accounts emphasize religious values, arguing either that certain Judaeo-Christian beliefs predispose adherents to devalue the environment (Schultz, Zelezny, & Dalrymple, 2000; White, 1967) or that beliefs that the environment is sacred enhance environmental concern (e.g., Dietz et al., 1998; Greeley, 1993; Kempton, Boster, & Hartley, 1995). Others have linked environmental concern and behavior to general theories of values (e.g., Schwartz, 1994) and have found that values those that focus concern beyond a person's immediate social circle (values called self-transcendent or altruistic) are stronger among people who engage in proenvironmental activities (e.g., Dietz et al., 1998; Karp, 1996; Stern & Dietz, 1994; Stern, Dietz, Kalof, & Guagnano, 1995). A related line of research finds greater evidence of environmental concern among individuals with "prosocial" rather than individualistic or competitive social value orientations (e.g., Joireman, Lasane, Bennett, Richards, & Solaimani, in press; Van Vugt & Samuelson, 1998).

Theories of altruistic behavior have also been used to explain environmentalism. This approach, first articulated by Heberlein (1972), presumes that because environmental quality is a public good, altruistic motives are a necessary for an individual to contribute to it in a significant way. The best developed example of this approach builds on Schwartz's (1973, 1977) moral norm-activation theory of altruism. The theory holds that altruistic (including proenvironmental) behavior occurs in response to personal moral norms that are activated in individuals who believe that particular conditions pose threats to others (awareness of adverse consequences, or AC) and that actions they could initiate could avert those consequences (ascription of responsibility to self, or AR). Substantial evidence supporting the theory's applicability to a range of environmental issues has accumulated over two decades (e.g., Black, 1978; Black et al., 1985; Guagnano, Stern, & Dietz, 1995; Schultz & Zelezny, 1999; Widegren, 1998).

My colleagues and I have developed a value-belief-norm (VBN) theory of environmentalism that builds on some of the above theoretical accounts and offers what we believe to be the best explanatory account to date of a variety of behavioral indicators of nonactivist environmentalism (Stern et al., 1999). The theory links value theory, norm-activation theory, and the New Environmental Paradigm (NEP) perspective through a causal chain of five variables leading to behavior: personal values (especially altruistic values), NEP, AC and AR beliefs about general conditions in the biophysical environment, and personal norms for proenvironmental action (see Figure 1). The rationale and empirical support for this causal ordering is drawn from previous work (Black et al., 1985; Gardner &

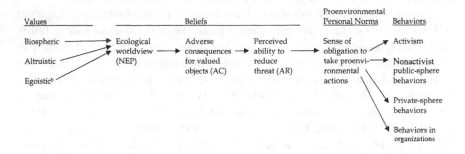

Fig. 1. A schematic representation of variables in the VBN theory of environmentalism[a]
[a]Arrows represent postulated direct effects. Direct effects may also be observed on variables more than one level downstream from a causal variable.
[b]Empirically, measures of egoistic values have been negatively correlated with indicators of environmentalism.

Stern, 1996; Stern, Dietz, & Guagnano, 1995; Stern, Dietz, Kalof, & Guagnano, 1995; Stern & Oskamp, 1987). The causal chain moves from relatively stable, central elements of personality and belief structure to more focused beliefs about human-environment relations (NEP), their consequences, and the individual's responsibility for taking corrective action. We postulate that each variable in the chain directly affects the next and may also directly affect variables farther down the chain. Personal norms to take proenvironmental action are activated by beliefs that environmental conditions threaten things the individual values (AC) and that the individual can act to reduce the threat (AR). Such norms create a general predisposition that influences all kinds of behavior taken with pro-environmental intent. In addition, behavior-specific personal norms and other social-psychological factors (e.g., perceived personal costs and benefits of action, beliefs about the efficacy of particular actions) may affect particular proenvironmental behaviors, as discussed below.

The VBN theory links value theory to norm-activation theory by generalizing the latter. It postulates that the consequences that matter in activating personal norms are adverse consequences to whatever the individual values (AC). Thus, people who value other species highly will be concerned about environmental conditions that threaten those valued objects, just as altruists who care about other people will be concerned about environmental conditions that threaten the other people's health or well-being. VBN theory links the NEP to norm-activation theory with the argument that the NEP is a sort of "folk" ecological theory from which beliefs about the adverse consequences of environmental changes can be deduced (for empirical support, see Stern, Dietz, & Guagnano, 1995).

In a recent study (Stern et al., 1999), my colleagues and I used the VBN theory, as well as measures from three other theories (indicators of four cultural biases, postmaterialist values, and belief in the sacredness of nature), to account for three types of nonactivist environmentalism: environmental citizenship, private-sphere behavior, and policy support (willingness to sacrifice). The VBN cluster of variables was a far stronger predictor of each behavioral indicator than the other theories, even when the other theories were taken in combination (see Table 1). None of the theories, however, was very successful in predicting the sole indicator of activism (participation in an environmental demonstration), which appears to depend on other factors in addition to an environmentalist predisposition.

The results provide strong initial support for the VBN theory's contentions that personal moral norms are the main basis for individuals' general predispositions to proenvironmental action (other studies supporting this conclusion include Bratt, 1999b, and Widegren, 1998) and that these norms are activated as the theory specifies. The personal norm variable was the only psychological variable of the 14 in the study that is associated with all three types of nonactivist environmentalism when the other variables are held constant. Moreover, values, NEP, and AC beliefs accounted for 56% of the variance in personal norms.

Table 1. Explained variance in Three Indicators of Proenvironmental Behavior

Source of explanatory variables	Dependent measures		
	Private-sphere behavior	Policy support	Environmental citizenship
VBN theory	.194	.346	.302
Three other theories[a]	.094	.199	.187
Added variance from other theories[b]	.033	.033	.091

Note. From "A Value-Belief-Norm Theory of Support for Social Movements: The Case of Environmental Concern," by P. C. Stern, T. Dietz, T. Abel, G. A. Guagnano, and L. Kalof, 1999, *Human Ecology Review*, 6, p. 90. Copyright 1999 by Society for Human Ecology. Reprinted with permission.
[a]Postmaterialist values, four cultural biases, and beliefs about the sacredness of nature.
[b]Difference between R^2 value for model combining VBN theory variables with the variables from the other three theories and value for model with VBN theory alone.

Data from several studies indicate that the values most strongly implicated in activating proenvironmental personal norms are, as norm-activation theory presumes, altruistic or self-transcendent values (Karp, 1996; Stern, Dietz, Kalof, & Guagnano, 1995; Stern et al., 1999). However, other values are sometimes linked as well. Self-enhancement or egoistic values and "traditional" values such as obedience, self-discipline, and family security are negatively associated with proenvironmental norms and action in some studies. The ways these values affect behavior are not well understood, but they may be important bases for principled opposition by some individuals to environmental movement goals. Another potentially important issue, as yet unresolved empirically, is whether a set of biospheric values is emerging, distinct from altruistic values about other people, that might provide a distinct basis for people's support for preserving endangered species and habitats.

An important element of the VBN theory is that the link from values to environmentalism is mediated by particular beliefs, such as beliefs about which kinds of people or things are affected by environmental conditions (AC) and about whether there are individual actions that could alleviate threats to valued persons or things (AR). Thus, environmentalist personal norms and the predisposition to proenvironmental action can be influenced by information that shapes these beliefs. This proposition suggests how environmentalism can be affected by the findings of environmental science (about consequences), publicity and commentary about those findings, and the actual and perceived openness of the political system to public influence (which may affect perceptions of personal responsibility). It also suggests an interpretation of environmentalist and antienvironmentalist rhetoric as efforts to activate or deactivate people's environmental norms by highlighting certain kinds of values or consequences (Stern, Dietz, Kalof, & Guagnano, 1995). The VBN theory offers an account of attitude formation that can deal with new or changing attitude objects (Stern, Dietz, Kalof, & Guagnano, 1995) and, more generally, with how environmental concern and environmental issues are socially constructed (Dietz, Stern, & Rycroft, 1989). The VBN theory is thus compatible with the constructed-preference tendency in cognitive psychology (Dietz & Stern, 1995; Fischhoff, 1991; Payne, Bettman, & Johnson, 1992).

The Causes of Environmentally Significant Behavior

Because environmental intent and environmental impact are two different things, theories explaining environmentalism are necessarily insufficient for understanding how to change environmentally important behaviors. Environmentalist intent is only one of the factors affecting behavior, and often, it is not one of the most important. Many environmentally significant behaviors are matters of personal habit or household routine (e.g., the setting of thermostats or the brand of paper towels purchased) and are rarely considered at all. Others are highly constrained by income or infrastructure (e.g., reinsulating homes, using public transport). For others, environmental factors are only minor influences on major actions (e.g., choosing an engine size option in a new automobile, deciding whether to centrally air condition a home), or the environmental effects are unknown to the consumer (e.g., choosing between products that have different environmental impacts from their manufacturing processes). Sometimes, as with spray cans, people may act in ways that are proenvironmental in intent but that in fact have little or no positive environmental impact. Environmentally beneficial actions may also follow from nonenvironmental concerns, such as a desire to save money, confirm a sense of personal competence, or preserve time for social relationships (De Young, this issue). And environmental concerns may fail to lead to proenvironmental action for various reasons (Gardner & Stern, 1996; Kempton, 1993). To understand any specific environmentally significant behavior requires empirical analysis. The evidence suggests that the role of environmentalist predispositions can vary greatly with the behavior, the actor, and the context.

ABC Theory

A first step toward understanding the complexities is to elaborate on the truism that behavior is a function of the organism and its environment. In one formulation (Guagnano et al., 1995), behavior (B) is an interactive product of personal-sphere attitudinal variables (A) and contextual factors (C). The attitude-behavior association is strongest when contextual factors are neutral and approaches zero when contextual forces are strongly positive or negative, effectively compelling or prohibiting the behavior in question (an inverted U-shaped function). We found supportive evidence for this formulation in a study of curbside recycling (Guagnano et al., 1995).

This "ABC theory" formulation implies that for personal behaviors that are not strongly favored by context (e.g., by being required or tangibly rewarded), the more difficult, time-consuming, or expensive the behavior, the weaker its dependence on attitudinal factors. Supporting evidence for this implication exists in studies that have used the same attitudinal variables to account for different proenvironmental behaviors. For example, in a study of household energy conservation, the relative explanatory power of social-psychological variables declined as effort or cost increased, from 59% of the explainable variance in self-reported home thermostat settings to 50% for minor curtailments such as shutting off heat in unused rooms, 44% for low-cost energy efficiency improvements such as caulking and weather-stripping, and 25% for major investments such as adding insulation or storm windows (Black et al., 1985). There are similar findings for public-sphere behaviors. The social-psychological variables of the VBN theory accounted for 35% of the variance in expressed policy support for environmentalism and 30% of the variance in environmental citizenship behaviors but only 4% of the variance in committed activism (Stern et al., 1999). These findings suggest a provocative hypothesis that is worthy of further exploration, namely that the more important a behavior is in terms of its environmental impact, the less it depends on attitudinal variables, including environmental concern.

Four Types of Causal Variables

It is useful to refine the personal-contextual or organism-environment distinction and to group the causal variables into four major types. *Attitudinal factors*, including norms, beliefs, and values, are one. The VBN theory provides a good theoretical account of one such factor, the general predisposition to act with proenvironmental intent, which can influence all behaviors an individual considers to be environmentally important. Other attitudinal variables affect only certain environmentally relevant behaviors. These include behavior-specific predispositions (e.g., specific personal moral norms in the terms of norm-activation theory, attitudes toward acts in the terms of the theory of planned behavior) and behavior-specific beliefs (e.g., about the difficulty of taking certain actions or about their consequences for self, others, or the environment). Several social-psychological theories, including cognitive dissonance theory, norm-activation theory, and the theory of planned behavior, have been shown to explain variance in specific proenvironmental behaviors. This research has demonstrated that pro-environmental behaviors can be affected by personal commitment and the perceived personal costs and benefits of particular actions (e.g., Katzev & Johnson, 1987) as well as by behavior-specific beliefs and personal norms (e.g., Black et al., 1985). As already noted, environmentally significant behavior can also be affected by nonenvironmental attitudes, such as those about attributes of consumer products that are correlated with environmental impact (e.g., speed, power, and luggage capacity in motor vehicles), or about frugality, luxury, waste, or the importance of spending time with family.

A second major type of causal variable is external or *contextual forces*. These include interpersonal influences (e.g., persuasion, modeling); community expectations; advertising; government regulations; other legal and institutional factors (e.g., contract restrictions on occupants of rental housing); monetary incentives and costs; the physical difficulty of specific actions; capabilities and constraints provided by technology and the built environment (e.g., building design, availability of bicycle paths, solar energy technology); the availability of public policies to support behavior (e.g., curbside recycling programs); and various features of the broad social, economic, and political context (e.g., the price of oil, the sensitivity of government to public and interest group pressures, interest rates in financial markets). It is worth nothing that a contextual factor may have different meanings to people with different attitudes or beliefs. For example, the higher price of "organic" produce may be an economic barrier to purchase for some people, whereas for others it is a marker of a superior product.

Personal capabilities are a third type of causal variable. These include the knowledge and skills required for particular actions (e.g., the skills of a movement organizer for activism, mechanical knowledge for energy-conserving home repairs), the availability of time to act, and general capabilities and resources such as literacy, money, and social status and power. Sociodemographic variables such as age, educational attainment, race, and income may be indicators or proxies for personal capabilities. Although these variables have very limited explanatory power for many environmentally significant behaviors (e.g., Dietz et al., 1998), they may be important for behaviors that depend strongly on particular capabilities. For instance, in a recent study (Stern et al., 1999), sociodemographic variables were found to be unrelated to consumer behavior and policy support when social-psychological variables were held constant, but environmental citizenship was found to be positively associated with income and with White race. The findings reflect the fact that the efficacy of environmental citizenship depends on an individual's social and economic resources. Also, environmental activism, for which attitudinal variables had very little explanatory power, was significantly associated (negatively) with age and income.

Finally, *habit or routine* is a distinct type of causal variable. Behavior change often requires breaking old habits and becomes established by creating new ones (Dahlstrand & Biel, 1997). Habit, in the form of standard operating procedure, is also a key factor in environmentally significant organizational behavior.

The evidence suggests that different types of causal variables are important, depending on the particular behavior (Gardner & Stern, 1996; Stern, 2000). Expensive behaviors such as reinsulating homes are likely to be strongly influenced by monetary factors; difficult behaviors such as reducing automobile use in the suburbs are likely to be strongly influenced by public policy supports (e.g., for alternative transport modes); behaviors that require specialized skills are likely to be strongly influenced by whether or not one possesses those capabilities; and so forth. Such hypotheses, though fairly obvious, do not go without saying. They offer a good starting point for efforts to understand particular environmentally significant behaviors.

Different causal variables also appear to work different ways in influencing behavior. For example, certain attitudinal factors create a general predisposition to act, which may be shaped into specific action largely by personal capabilities and contextual forces. A new context may make old habits untenable and lead someone to consider his or her attitudes and values explicitly in developing new ones (Dahlstrand & Biel, 1997). Or financial incentives may favor behaviors that nevertheless do not occur unless information makes individuals aware that the incentive is available (Stern, 1999).

The insight of the ABC formulation, that the different types of causal factors may interact, implies that interpretations based only on main effects can be seriously misleading. Studies that examine only attitudinal factors are likely to find effects only inconsistently, because the effects are contingent on capabilities and context. Similarly, studies that examine only contextual variables, such as material incentives, social norms, or the introduction of new technology, may find effects but fail to reveal their dependence on individuals' attitudes or beliefs. Single-variable studies may demonstrate that a particular theoretical framework has explanatory power but may not contribute much to the comprehensive understanding of particular environmentally significant behaviors that is needed to change them. I return to this point later.

Toward a Synthesis

The field now needs synthetic theories or models that incorporate variables from more than one of the above broad classes, postulate relationships among them, and use them to explain one or more types of environmentally significant behavior. Researchers are beginning to propose such models (e.g., Dahlstrand & Biel, 1997; Fransson & Gärling, 1999; Gardner & Stern, 1996; Hines, Hungerford, & Tomera, 1987; Ölander & Thøgerson, 1995; Stern & Oskamp, 1987; Vlek, 2000). Some of the models expand on familiar theories of altruistic behavior (e.g., Schwartz, 1977) or planned behavior (e.g., Ajzen, 1991), which emphasize attitudinal factors almost exclusively. Because the new models also take into account personal capabilities, context, and habits, they are more suitable for explaining behaviors that have significant environmental impacts, which are often strongly influenced by such nonattitudinal factors.

A dialogue among such models is needed to move the field toward synthesis. It is also likely to build links to other psychological theories. For example, the distinction between attitudes and habits as causes of behavior closely parallels the distinction in a variety of "dual-process" models (Smith & DeCoster, 2000) between conscious and effortful behaviors and automatic or associative ones. Dual-process models may therefore have something to say about pro-environmental behavior.

Changing Environmentally Significant Behavior

Many approaches toward changing individuals' environmentally significant behavior have been tried. Gardner and Stern (1996) reviewed the evidence on four major types of intervention: religious and moral approaches that appeal to values and aim to change broad worldviews and beliefs; education to change attitudes and provide information; efforts to change the material incentive structure of behavior by providing monetary and other types of rewards or penalties; and community management, involving the establishment of shared rules and expectations. They found that each of these intervention types, if carefully executed, can change behavior. However, moral and educational approaches have generally disappointing track records, and even incentive- and community-based approaches rarely produce much change on their own. By far, the most effective behavior change programs involve combinations of intervention types.

These findings underline the limits of single-variable explanations for informing efforts at behavior change. The behavior is determined by multiple variables, sometimes in interaction. There is strong evidence, for example, that incentives and information interact, with the combination sometimes being much more effective than the sum of the two interventions (Stern, 1999). In one evaluation study, increased financial incentives for major investments in home energy conservation were necessary but far from sufficient for programs to be successful. Even when electric utility companies offered to subsidize 93% of the cost of home insulation, consumer response varied from 1% to almost 20% adoption per year, apparently depending on how the subsidy was made known to householders (Stern et al., 1986).

Often the nature of the interaction can be well described in terms of barriers or limiting conditions to behavior change (Gardner & Stern, 1996). Interventions do little or nothing until one of them removes an important barrier to change. To promote investments in home insulation, for example, it is necessary to reduce the financial barriers, provide accurate information on which actions would be effective, and reduce the difficulty of getting the information and finding a reliable contractor. Programs that did all these things were vastly more successful than programs that did only one or two (Stern et al., 1986). Since different individuals face different impediments to behavior change and the impediments are often multiple, little happens until the right combination of intervention types is found. The concept of limiting conditions also implies that particular kinds of interventions have diminishing returns after they have fulfilled their major function. For example, once financial incentives are large enough to demonstrate a clear personal benefit, increasing the incentive may be far less effective in producing behavior change than providing information through marketing (see Stern, 1999).

Theory has progressed to the point at which it is possible to identify useful and practical principles for intervention (see Table 2; for a guide to the application of these principles, see McKenzie-Mohr & Smith, 1999). Space does not permit elaboration of all the principles here. The admonitions to combine multiple intervention types, to understand the situation from the actor's perspective, to continually monitor and adjust programs, and to use participatory methods all suggest ways to make practical progress with incomplete theory.

For researchers who would like to advance the understanding necessary to make behavioral approaches to environmental protection more successful, a related set of principles applies (see Gardner & Stern, 1996, chap. 10). First, identify target behaviors that are environmentally significant in terms of impact. Then analyze the behaviors to identify the responsible actors and actions. Then consider the full range of causal variables and explore their possible relevance to the target behavior from the actor's standpoint. By exploring the possibilities directly with representatives of the population whose behavior is to be changed, it is possible to find promising strategies for intervention without trying them all out experimentally.

This research strategy offers the best approach to developing useful theory about specific behavioral types that have important environmental impacts. In addition to its practical value, such small-scale theory provides the essential building blocks for broader, inductively developed theory about environmentally significant behavior.

Table 2. Principles for Intervening to Change Environmentally Destructive Behavior

A. Use multiple intervention types to address the factors limiting behavior change

    1.   Limiting factors are numerous (e.g., technology, attitudes, knowledge, money, convenience, trust)

    2.   Limiting factors vary with actor and situation, and over time

    3.   Limiting factors affect each other

B. Understand the situation from the actor's perspective

C. When limiting factors are psychological, apply understanding of human choice processes

    1.   Get the actors' attention; make limited cognitive demands

    2.   Apply principles of community management (credibility, commitment, face-to-face communication, etc.)

D. Address conditions beyond the individual that constrain proenvironmental choice

E. Set realistic expectations about outcomes

F. Continually monitor responses and adjust programs accordingly

G. Stay within the bounds of actors' tolerance for intervention

H. Use participatory methods of decision making

Note. From *Environmental Problems and Human Behavior* (p. 159), by G. T. Gardner and P. C. Stern, 1996, Boston: Allyn and Bacon. Copyright 1996 by Allyn and Bacon. Reprinted with permission.

Conclusions

Environmentally significant behavior is dauntingly complex, both in its variety and in the causal influences on it. Although a general theory lies far in the distance, enough is known to present a framework that can increase theoretical coherence. This framework includes typologies of environmentally significant behaviors and of their causes (see Table 3) and a growing set of empirical propositions about these variables. For example:

- The VBN approach offers a good account of the causes of the general predisposition toward proenvironmental behavior.

- Environmentally significant behavior depends on a broad range of causal factors, both general and behavior-specific. A general theory of environmentalism may therefore not be very useful for changing specific behaviors.

- Different kinds of environmentally significant behavior have different causes. Because the important causal factors may vary greatly across behaviors and individuals, each target behavior should be theorized separately.

Table 3. Major Types of Environmentally Significant Behaviors and Causal Variables Influencing These Behaviors

Causal variables	Environmentally significant behaviors
Attitudinal	*Environmental activism*
General environmentalist predisposition[a]	*Nonactivist public-sphere behaviors*
Behavior-specific norms and beliefs[b]	Environmental citizenship
Nonenvironmental attitudes	(e.g., petitioning, joining groups)
(e.g., about product attributes)	Policy support
Perceived costs and benefits of action	
Personal capabilities	*Private-sphere environmentalism*
Literacy	Consumer purchase behaviors
Social status	Maintenance of household equipment
Financial resources	Changes in equipment use, lifestyle (curtailment)
Behavior-specific knowledge and skills	Waste disposal behaviors
	"Green consumerism"
Contextual factors	*Other*
Material costs and rewards	Behaviors affecting organizational decisions
Laws and regulations	
Available technology	
Social norms and expectations	
Supportive policies	
Advertising	
Habit and routine	

[a]The VBN theory incorporates various attitudinal variables believed to create this predisposition.
[b]These norms and beliefs figure prominently in applications of norm-activation theory and the theory of planned behavior to specific proenvironmental behaviors.

- The causal factors may interact. Attitudinal causes have the greatest predictive value for behaviors that are not strongly constrained by context or personal capabilities. For behaviors that are expensive or difficult, contextual factors and personal capabilities are likely to account for more of the variance.

In addition to such empirical principles, past research has yielded important insights for research and action on environmental protection, as described above and in Table 2. One cannot overemphasize to behavioral scientists the importance of identifying target behaviors from an environmental perspective (in terms of their impact), even though understanding them requires an actor-oriented approach that focuses on their causes. It is also critical to underscore the need to draw on insights from across the behavioral and social sciences, because the important causal variables lie in the domains of various disciplines and because the variables interact. Thus, interdisciplinary research is necessary for full understanding.

By following these insights and elaborating on the above principles, behavioral researchers can further advance understanding of environmentally significant individual behavior and can provide useful input to practical programs for environmental protection. They are also likely to make contributions to the broader project of behavioral science.

References

Ajzen, I. (1991). The theory of planned behavior. *Organizational Decision and Human Decision Process*, *50*, 179–211.

Allen, J. B., & Ferrand, J. L. (1999). Environmental locus of control, sympathy, and proenvironmental behavior: A test of Geller's actively caring hypothesis. *Environment and Behavior*, *31*, 338–353.

Black, J. S. (1978). Attitudinal, normative, and economic factors in early response to an energy-use field experiment (Doctoral dissertation, University of Wisconsin, 1978). *Dissertation Abstracts International*, *39*, 436B.

Black, J. S., Stern, P. C., & Elworth, J. T. (1985). Personal and contextual influences on household energy adaptations. *Journal of Applied Psychology*, *70*, 3–21.

Bratt, C. (1999a). Consumers' environmental behavior: Generalized, sector-based, or compensatory? *Environment and Behavior*, *31*, 28–44.

Bratt, C. (1999b). The impact of norms and assumed consequences on recycling behavior. *Environment and Behavior*, *31*, 630–656.

Dahlstrand, U., & Biel, A. (1997). Pro-environmental habits: Propensity levels in behavioral change. *Journal of Applied Social Psychology*, *27*, 588–601.

Dake, K. (1991). Orienting dispositions in the perception of risk: An analysis of contemporary worldviews and cultural biases. *Journal of Cross-Cultural Psychology*, *22*, 61–82.

Dietz, T., & Stern, P. C. (1995). Toward a theory of choice: Socially embedded preference construction. *Journal of Socio-Economics*, *24*, 261–279.

Dietz, T., Stern, P. C., & Guagnano, G. A. (1998). Social structural and social psychological bases of environmental concern. *Environment and Behavior*, *30*, 450–471.

Dietz, T., Stern, P. C., & Rycroft, R. W. (1989). Definitions of conflict and the legitimation of resources: The case of environmental risk. *Sociological Forum*, *4*, 47–70.

Douglas, M., & Wildavsky, A. (1982). *Risk and culture: An essay on the selection of technological and environmental dangers*. Berkeley and Los Angeles: University of California Press.

Fischhoff, B. (1991). Preference elicitation: Is there anything in there? *American Psychologist, 46,* 835–847.

Fransson, N., & Gärling, T. (1999). Environmental concern: Conceptual definitions, measurement methods, and research findings. *Journal of Environmental Psychology, 19,* 369–382.

Gardner, G. T., & Stern, P. C. (1996). *Environmental problems and human behavior.* Boston: Allyn and Bacon.

Greeley, A. (1993). Religion and attitudes toward the environment. *Journal for the Scientific Study of Religion, 32,* 19–28.

Guagnano, G. A., Stern, P. C., & Dietz, T. (1995). Influences on attitude-behavior relationships: A natural experiment with curbside recycling. *Environment and Behavior, 27,* 699–718.

Heberlein, T. A. (1972). The land ethic realized: Some social psychological explanations for changing environmental attitudes. *Journal of Social Issues, 28*(4), 79–87.

Hines, J. M., Hungerford, H. R., & Tomera, A. N. (1987). Analysis and synthesis of research on responsible environmental behavior: A meta-analysis. *Journal of Environmental Education, 18,* 1–18.

Inglehart, R. (1990). *Culture shift in advanced industrial society.* Princeton, NJ: Princeton University Press.

Joireman, J. A., Lasane, T. P., Bennett, J., Richards, D., & Solaimani, S. (in press). Integrating social value orientation and the consideration of future consequences within the extended norm activation model of proenvironmental behavior. *British Journal of Social Psychology.*

Kals, E., Schumacher, D., & Montada, L. (1999). Emotional affinity toward nature as a motivational basis to protect nature. *Environment and Behavior, 31,* 178–202.

Karp, D. G. (1996). Values and their effects on pro-environmental behavior. *Environment and Behavior, 28,* 111–133.

Katzev, R. D., & Johnson, T. R. (1987). *Promoting energy conservation: An analysis of behavioral techniques.* Boulder, CO: Westview Press.

Kempton, W. (1993). Will public environmental concern lead to action on global warming? *Annual Review of Energy and Environment, 18,* 217–245.

Kempton, W., Boster, J. S., & Hartley, J. A. (1995). *Environmental values in American culture.* Cambridge, MA: MIT Press.

McAdam, D., McCarthy, J. D., & Zald, M. N. (1988). Social movements. In N. J. Smelser (Ed.), *Handbook of sociology* (pp. 695–738). Newbury Park, CA: Sage.

McKenzie-Mohr, D., & Smith, W. (1999). *Fostering sustainable behavior: An introduction to community-based social marketing.* Gabriola Island, British Columbia, Canada: New Society Publishers.

Ölander, F., & Thøgerson, J. (1995). Understanding consumer behavior as a prerequisite for environmental protection. *Journal of Consumer Policy, 18,* 345–385.

Payne, J. W., Bettman, J. R., & Johnson, E. J. (1992). Behavioral decision research: A constructive processing perspective. *Annual Review of Psychology, 43,* 87–131.

Rosa, E. A., & Dietz, T. (1998). Climate change and society: Speculation, construction and scientific investigation. *International Sociology, 13,* 421–425.

Schultz, P. W., & Zelezny, L. C. (1999). Values as predictors of environmental attitudes: Evidence for consistency across cultures. *Journal of Environmental Psychology, 19,* 255–265.

Schultz, P. W., Zelezny, L. C., & Dalrymple, N. J. (2000). A multinational perspective on the relation between Judeo-Christian religious beliefs and attitudes of environmental concern. *Environment and Behavior, 32,* 576–591.

Schwartz, S. H. (1973). Normative explanations of helping behavior: A critique, proposal, and empirical test. *Journal of Experimental Social Psychology, 9,* 349–364.

Schwartz, S. H. (1977). Normative influences on altruism. In L. Berkowitz (ed.), *Advances in experimental social psychology* (Vol. 10, pp. 221–279). New York: Academic Press.

Schwartz, S. H. (1994). Are there universal aspects in the structure and contents of human values? *Journal of Social Issues, 50*(4), 19–46.

Smith, E. R., & DeCoster, J. (2000). Dual-process models in social and cognitive psychology: Conceptual integration and links to underlying memory systems. *Personality and Social Psychology Review, 4,* 108–131.

Steg, L., & Sievers, I. (2000). Cultural theory and individual perceptions of environmental risks. *Environment and Behavior, 332,* 250–269.

Stern, P. C. (1997). Toward a working definition of consumption for environmental research and policy. In P. C. Stern, T. Dietz, V. R. Ruttan, R. H. Socolow, & J. L. Sweeney (Eds.), *Environmentally significant consumption: Research directions* (pp. 12–35). Washington, DC: National Academy Press, 1997.

Stern, P. C. (1999). Information, incentives, and proenvironmental consumer behavior. *Journal of Consumer Policy, 22*, 461–478.

Stern, P. C. (2000). Psychology, sustainability, and the science of human-environment interactions. *American Psychologist, 55*, 523–530.

Stern, P. C., Aronson, E., Darley, J. M., Hill, D. H., Hirst, E., Kempton, W., & Wilbanks, T. J. (1986). The effectiveness of incentives for residential energy conservation. *Evaluation Review, 10*(2), 147–176.

Stern, P. C., & Dietz, T. (1994). The value basis of environmental concern. *Journal of Social Issues, 50*(3), 65–84.

Stern, P. C., Dietz, T., Abel, T., Guagnano, G. A., & Kalof, L. (1999). A value-belief-norm theory of support for social movements: The case of environmental concern. *Human Ecology Review, 6*, 81–97.

Stern, P. C., Dietz, T., & Guagnano, G. A. (1995). The new environmental paradigm in social psychological perspective. *Environment and Behavior, 27*, 723–745.

Stern, P. C., Dietz, T., Kalof, L., & Guagnano, G. A. (1995). Values, beliefs and proenvironmental action: Attitude formation toward emergent attitude objects. *Journal of Applied Social Psychology, 25*, 1611–1636.

Stern, P. C., & Gardner, G. T. (1981a). Psychological research and energy policy. *American Psychologist 36*, 329–342.

Stern, P. C., & Gardner, G. T. (1981b). The place of behavior change in managing environmental problems. *Zeitschrift für Umweltpolitik, 2*, 213–239.

Stern, P. C., & Oskamp, S. (1987). Managing scarce environmental resources. In D. Stokols & I. Altman (Eds.), *Handbook of environmental psychology* (pp. 1043–1088). New York: Wiley.

Stern, P. C., Young, O. R., & Druckman, D. (Eds.). (1992). *Global environmental change: Understanding the human dimensions*. Washington, DC: National Academy Press.

Van Vugt, M., & Samuelson, C. D. (1998). The impact of personal metering in the management of a natural resource crisis: A social dilemma analysis. *Personality and Social Psychology Bulletin, 25*, 731–745.

Vayda, A. P. (1988). Actions and consequences as objects of explanation in human ecology. In R. J. Borden, J. Jacobs, & G. L. Young (Eds.), *Human ecology: Research and applications* (pp. 9–18). College Park, MD: Society for Human Ecology.

Vlek, C. (2000). Essential psychology for environmental policy making. *International Journal of Psychology, 35*, 153–167.

White, L., Jr. (1967). The historical roots of our ecological crisis. *Science, 155*, 1203–1207.

Widegren, Ö. (1998). The new environmental paradigm and personal norms. *Environment and Behavior, 30*, 75–100.

Zald, M. (1992). Looking backward to look forward: Reflections on the past and future of the resource mobilization research program. In A. D. Morris & C. M. Mueller (Eds.), *Frontiers in social movement theory* (pp. 326–348). New Haven, CT: Yale University Press.

PAUL C. STERN is Study Director of the Committee on the Human Dimensions of Global Change at the U.S. National Research Council. He is also a Research Professor of Sociology at George Mason University and President of the Social and Environmental Research Institute. His current research interests include the study of environmental values, beliefs, and behavior and the development of deliberative approaches to environmental decision making. Recent publications include the coedited volumes *Understanding Risk: Informing Decisions in a Democratic Society* (1996), *Environmentally Significant Consumption: Research Directions* (1997), *Making Climate Forecasts Matter* (1999), and *International Conflict Resolution After the Cold War* (2000), all published by National Academy Press.

Journal of Social Issues, Vol. 56, No. 3, 2000, pp. 425–442

Measuring Endorsement of the New Ecological Paradigm: A Revised NEP Scale

Riley E. Dunlap*
Washington State University

Kent D. Van Liere
Primen

Angela G. Mertig
Michigan State University

Robert Emmet Jones
University of Tennessee

Dunlap and Van Liere's New Environmental Paradigm (NEP) Scale, published in 1978, has become a widely used measure of proenvironmental orientation. This article develops a revised NEP Scale designed to improve upon the original one in several respects: (1) It taps a wider range of facets of an ecological worldview, (2) It offers a balanced set of pro- and anti-NEP items, and (3) It avoids outmoded terminology. The new scale, termed the New Ecological Paradigm Scale, consists of 15 items. Results of a 1990 Washington State survey suggest that the items can be treated as an internally consistent summated rating scale and also indicate a modest growth in pro-NEP responses among Washington residents over the 14 years since the original study.

When environmental issues achieved a prominent position on our nation's policy agenda in the 1970s, the major problems receiving attention tended to be air

*This is a revision of a paper presented at the Annual Meeting of the Rural Sociological Society, The Pennsylvania State University, State College, PA, August 1992. Revision of the NEP Scale benefited from Dunlap's long-term collaborative effort with William R. Catton, Jr., to document the emergence of an ecological paradigm within sociology. The data reported in this article were collected in a survey sponsored by Washington State University's Department of Natural Resource Sciences and Cooperative Extension Service for which Dunlap served as a consultant. Thanks are extended to Robert Howell for facilitating Dunlap's involvement with that survey. Correspondence concerning this article should be addressed to Riley E. Dunlap, Department of Sociology, Washington State University, Pullman, WA 99164-4020 [e-mail: dunlap@wsu.edu].

425

and water pollution, loss of aesthetic values, and resource (especially energy) conservation. Consequently, attempts to measure public concern for environmental quality, or "environmental concern," focused primarily on such conditions (e.g., Weigel & Weigel, 1978). In recent decades, however, environmental problems have evolved in significant ways. Although localized pollution, especially hazardous waste, continues to be a major issue, environmental problems have generally tended to become more geographically dispersed, less directly observable, and more ambiguous in origin. Not only do problems such as ozone depletion, deforestation, loss of biodiversity, and climate change cover far wider geographical areas (often reaching the global level), but their causes are complex and synergistic and their solutions complicated and problematic (Stern, Young, & Druckman, 1992). Researchers interested in understanding how the public sees environmental problems are gradually paying attention to these newly emerging "attitude objects" (Stern, Dietz, Kalof, & Guagnano, 1995), and the number of studies of public perceptions of issues such as global warming is slowly mounting (Dunlap, 1998; O'Connor, Bord, & Fisher, 1999).

The emergence of global environmental problems as major policy issues symbolizes the growing awareness of the problematic relationship between modern industrialized societies and the physical environments on which they depend (Stern et al., 1992). Recognition that human activities are altering the ecosystems on which our existence—and that of all other living species—is dependent and growing acknowledgment of the necessity of achieving more sustainable forms of development give credence to suggestions that we are in the midst of a fundamental reevaluation of the underlying worldview that has guided our relationship to the physical environment (e.g., Milbrath, 1984). In particular, suggestions that a more ecologically sound worldview is emerging have gained credibility in the past decade (e.g., Olsen, Lodwick, & Dunlap, 1992).

In this context, it is not surprising to see that traditional measures of "environmental concern" are being supplanted by instruments seeking to measure "ecological consciousness" (Ellis & Thompson, 1997), "anthropocentrism" (Chandler & Dreger, 1993), and "anthropocentrism versus ecocentrism" (Thompson & Barton, 1994). The purpose of this article is to provide a revision of the earliest such measure of endorsement of an ecological worldview, the New Environmental Paradigm Scale (Dunlap & Van Liere, 1978).

The New *Environmental* Paradigm Scale

Development of the Scale

Sensing that environmentalists were calling for more far-reaching changes than the development of environmental protection policies and stimulated by Pirages and Ehrlich's (1974) explication of the antienvironmental thrust of our society's dominant social paradigm (DSP), in the mid-1970s Dunlap and Van Liere argued that implicit within environmentalism was a challenge to our fundamental views about nature and humans' relationship to it. Their conceptualization of what they called the New Environmental Paradigm (NEP) focused on beliefs about humanity's ability to upset the balance of nature, the existence of limits to growth for human societies, and humanity's right to rule over the rest of nature. In a 1976 Washington State study Dunlap and Van Liere (1978) found that a set of 12 Likert items measuring these three facets of the new social paradigm or worldview exhibited a good deal of internal consistency (coefficient alpha of .81), and strongly discriminated between known environmentalists and the general public. Consequently, they argued that the items could legitimately be treated as a New Environmental Paradigm Scale, and found that endorsement of the NEP was, as expected, negatively related to endorsement of the DSP (Dunlap & Van Liere, 1984). [Dunlap and Van Liere later developed a six-item NEP Scale for use in a national survey for the Continental Group (1982) that has subsequently been used by several researchers, particularly political scientists (Pierce, Steger, Steel, & Lovrich, 1992).]

Drawing upon a spate of literature in the late 1970s and early 1980s that explicated more fully the contrast between the emerging environmental paradigm and the dominant social paradigm (e.g., Brown, 1981), subsequent researchers provided far more comprehensive conceptualizations of the NEP and DSP (Cotgrove, 1982; Milbrath, 1984; Olsen et al., 1992). However, their elaborate measuring instruments, encompassing a wide range of both beliefs and values, have proven unwieldy, and the NEP Scale has become the far more widely used measure of an environmental or, as now seems the more appropriate label, "ecological" worldview. Also, because the emergence of global environmental change has made items like "The balance of nature is very delicate and easily upset" more relevant now than in the 1970s, and because alternative measures of environmental concern widely used in the 1970s and early 1980s focusing on specific types of environmental problems have become dated (e.g., Weigel & Weigel, 1978), the NEP Scale has also become a popular measure of environmental concern, with endorsement of the NEP treated as reflecting a proenvironmental orientation.

The fact that the NEP Scale is treated as a measure of endorsement of a fundamental paradigm or worldview, as well as of environmental attitudes, beliefs, and even values, reflects the ambiguity inherent in measuring these phenomena as well as Dunlap and Van Liere's failure to ground the NEP in social-psychological theories of attitude structure (Stern, Dietz, & Guagnano, 1995). Although attitude theory cautions against categorizing individual items as clear-cut indicators of attitudes *or* beliefs (see, e.g., Eagly & Kulesa, 1997), in retrospect it nonetheless seems reasonable to argue that the NEP items primarily tap "primitive beliefs" about the nature of the earth and humanity's relationship with it. According to Rokeach (1968, p. 6), primitive beliefs form the inner core of a person's belief system and "represent his 'basic truths' about physical reality, social reality and the nature of the self." Though not as foundational as the examples used by Rokeach, beliefs about nature and humans' role in it as measured by the NEP items appear to constitute a fundamental component of people's belief systems vis-à-vis the environment.

Social psychologists see these primitive beliefs as influencing a wide range of beliefs and attitudes concerning more specific environmental issues (see Gray, 1985, chap. 2, and Stern, Dietz, & Guagnano, 1995a, for two alternative but complementary models incorporating the NEP as a measure of primitive beliefs). Similarly, political scientists find the NEP beliefs to be a core element in comprehensive environmental belief systems (Dalton, Gontmacher, Lovrich, & Pierce, 1999; Pierce, Lovrich, Tsurutani, & Takematsu, 1987). A consensus that the NEP items measure such beliefs (Edgell & Nowell, 1989; Gooch, 1995) is emerging, and it seems reasonable to regard a coherent set of these beliefs as constituting a paradigm or worldview that influences attitudes and beliefs toward more specific environmental issues (Dalton et al., 1999). In short, a proecological orientation or "seeing the world ecologically," reflected by a high score on the NEP Scale, should lead to proenvironmental beliefs and attitudes on a wide range of issues (Pierce, Dalton, & Zaitsev, 1999; Stern, Dietz, & Guagnano, 1995). Although such beliefs may also influence behavior, the barriers and opportunities that influence proenvironmental behaviors in specific situations caution against expecting a strong NEP-behavior relationship (Gardner & Stern, 1996).

Past Research and Validity of the NEP Scale

Although treated variously as measuring environmental attitudes, beliefs, values, and worldview, the NEP Scale has been widely used during the past 2 decades. It has been used most often with samples of the general public, but it has also been used with samples of specific sectors such as farmers (Albrecht, Bultena, Hoiberg, & Nowak, 1982) and members of interest groups (e.g., Edgell & Nowell, 1989; Pierce et al., 1992). It has been used as well to examine the environmental orientations of ethnic minorities in the United States (e.g., Caron, 1989; Noe & Snow, 1989–90) as well as of residents of other nations such as Canada (Edgell & Nowell, 1989), Sweden (Widegren, 1998), the Baltic states (Gooch, 1995), Turkey (Furman, 1998), and Japan (Pierce et al., 1987). Finally, it has recently been used to compare the environmental orientations of college students in several Latin American nations and Spain with those of American students (Bechtel, Verdugo, & Pinheiro, 1999; Schultz & Zelezny, 1998). In general, these studies have found, as did Dunlap and Van Liere (1978) in their 1976 Washington State survey, a relatively strong endorsement of NEP beliefs across the various samples.

Rather than attempt to summarize the dozens of studies that have employed the NEP items, we will cite selected findings that bear on the validity of the NEP

Scale. As noted previously, studies of interest groups such as environmental organizations have consistently found that environmentalists score higher on the NEP
Scale than do the general public or members of nonenvironmental interest groups
(e.g., Edgell & Nowell, 1989; Pierce et al., 1992; Widegren, 1998). These findings
suggest, as did Dunlap and Van Liere's (1978) original study, that the NEP Scale
has known-group validity. Similarly, despite the difficulty of predicting behaviors
from general attitudes and beliefs, numerous studies have found significant relationships between the NEP Scale and various types of behavioral intentions as well
as both self-reported and observed behaviors (e.g., Blake, Guppy, & Urmetzer,
1997; Ebreo, Hershey, & Vining, 1999; O'Connor et al., 1999; Roberts & Bacon,
1997; Schultz & Oskamp, 1996; Schultz & Zelezny, 1998; Scott & Willits, 1994;
Stern, Dietz, & Guagnano, 1995a; Tarrant & Cordell, 1997; Vining & Ebreo,
1992). Such findings clearly indicate that the NEP Scale possesses predictive
validity as well. Since both predictive and known-group validity are forms of criterion validity (Zeller & Carmines, 1980, pp. 79–81), the overall evidence thus suggests that the NEP possesses criterion validity.

Judging the content validity of the NEP Scale is more difficult, especially
since the construct of an environmental/ecological paradigm or worldview is
inherently somewhat amorphous. A recent study by Kempton, Boster, and Hartley
(1995), however, that employed in-depth, ethnographic interviews in an effort to
flesh out the environmental perspectives of Americans is highly relevant in this
regard. Although their methods were dramatically different than those employed in
the development and construction of the NEP Scale, on the basis of responses to
their unstructured interviews Kempton et al. (1995, chap. 3) concluded that three
general sets of environmental beliefs play crucial roles in the "cultural models" by
which Americans attempt to make sense of environmental issues: (1) Nature is a
limited resource upon which humans rely; (2) Nature is balanced, highly interdependent and complex, and therefore susceptible to human interference; and
(3) Materialism and lack of contact with nature have led our society to devalue
nature. That Kempton et al. found three nearly identical beliefs to those forming the
major facets of the NEP Scale—balance of nature, limits to growth, and human
domination over nature—is strong confirmation of the scale's content validity.

Judging the construct validity of measuring instruments is difficult because it
depends on how the measure relates to other measures in ways that are theoretically
specified (Zeller & Carmines, 1980, pp. 80–84). Original claims of the NEP
Scale's construct validity (Dunlap & Van Liere, 1978, p. 16) were limited to the
fact that scores on it were related in the expected fashion with personal characteristics such as age (younger people were assumed to be less wedded to traditional
worldviews and thus more supportive of the NEP), education (the better educated
were assumed to be exposed to more information about environmental issues and
to be more capable of comprehending the ecological perspective implicit in the
NEP) and political ideology (liberals were assumed to be less committed to the

status quo in general and the DSP in particular). Although there have been some exceptions, most studies have continued to find support for the NEP to be negatively related to age and positively related to education and liberalism.

More importantly, studies that have examined the presumed intervening links between these variables and support for the NEP, such as those that have documented the assumed positive relationship between environmental knowledge and endorsement of the NEP (Arcury, 1990; Arcury, Johnson, & Scollay, 1986; Furman, 1998; Pierce et al., 1992) and two that found a negative relationship between right-wing authoritarianism and support for the NEP (Lefcourt, 1996; Schultz & Stone, 1994), are beginning to provide more convincing evidence of the NEP Scale's construct validity. But the most important evidence of the NEP Scale's construct validity comes from studies that have theorized that the NEP forms a primary component, along with fundamental values, of environmental belief systems and then have found this expectation empirically confirmed (Pierce et al., 1987; Stern, Dietz, & Guagnano, 1995). As theoretical models of the sources of environmental attitudes and behaviors that assign a key role to the NEP are developed, tested, and confirmed, evidence of the NEP Scale's construct validity should increase.

Dimensionality of the NEP Scale

While the bulk of available evidence converges to suggest the overall validity of the NEP Scale, there is far less consensus on the question of whether the scale measures a single construct or is inherently multidimensional. After a series of U.S. studies (Albrecht et al., 1982; Geller & Lasley, 1985; Noe & Snow, 1990) produced similar results via factor analysis, suggesting that the NEP is composed of three distinct dimensions—balance of nature, limits to growth, and human domination of nature—some researchers began to routinely measure each dimension separately (e.g., Arcury, 1990; Ebreo et al., 1999; Vining & Ebreo, 1992). A careful review of studies that have factor-analyzed the NEP items, however, reveals considerable inconsistency in the number of dimensions actually obtained: Three studies (Edgell & Nowell, 1989; Lefcourt, 1996; Noe & Snow, 1990, p. 24) found all items to load on a single factor with at least one of their samples, and several studies have found only two dimensions in one or more of their samples (Bechtel et al., 1999; Gooch, 1995; Noe & Snow, 1989–90, 1990; Noe & Hammitt, 1992; Scott & Willits, 1994). Although a number of studies have found three dimensions similar to those noted above in one or more samples (Edgell & Nowell, 1989; Noe & Snow, 1989–90; Shetzer, Stackman, & Moore, 1991), still others have found as many as four dimensions (Furman, 1998; Roberts & Bacon, 1997).

The above results, combined with the fact that studies finding three dimensions often report some discrepancies in the loadings of individual items, suggest that it may be premature to assume automatically that the 12 NEP items measure three distinct dimensions. We encourage researchers to at least factor-analyze the entire set of items at the outset to determine if the three widely used dimensions do in fact emerge. Factor-analyzing 12 items typically yields two or more dimensions, but as the above results indicate, the dimensions are often sample specific. For this reason, some researchers see unidimensionality as an unrealistic goal and settle for a high level of internal consistency, as measured by strong item-total correlations, high loadings on the first *unrotated* factor, and an acceptable (0.7 or higher) value for coefficient alpha, the mean of all possible split-half reliabilities (Zeller & Carmines, 1980, chap. 3). Although internal consistency is generally a necessary but not a sufficient condition for unidimensionality, it provides a reasonable rationale for combining a set of items into a single measure rather than creating ad hoc dimensions that emerge from various factoring techniques.

The decision to break the NEP items into two or more dimensions should depend upon the results of the individual study. If two or more distinct dimensions that have face validity emerge and are not highly correlated with one another, then it is sensible to employ them as separate variables. If substantively meaningful dimensions do not emerge, however, and the entire set of items (or at least a majority of them) are found to produce an internally consistent measure, then we recommend treating the NEP Scale as a single variable. Although the notion of a worldview or paradigm implies some consistency (in terms of taking pro- or anti-NEP positions) in responses to the NEP items, it is not unreasonable to expect that discernible dimensions will emerge in some samples, as populations vary in terms of how well their belief systems are organized into coherent frameworks (e.g., Bechtel et al. 1999; Dalton et al., 1999; Gooch, 1995; Pierce et al., 1987). Thus, the decision to treat the NEP as a single variable or as multiple variables should not be made beforehand but ought to be based on the results of the particular study. Whether used as a single scale or as a multidimensional measure, the NEP can still be fruitfully employed to examine the structure and coherence of ecological worldviews and the relationships between these worldviews and a range of more specific environmental attitudes, beliefs, and behaviors.

Finally, it should also be noted that the apparent multidimensionality of the NEP items may stem in part from a serious flaw in the original 12-item NEP Scale. Only 4 of the 12 items were worded in an anti-NEP direction, *and all 4 focused on anthropocentrism or the belief that nature exists primarily for human use and has no inherent value of its own.* That these items generally form a distinct dimension (often termed "domination of nature") in factor-analytic studies reporting two or more dimensions may thus represent a methodological artifact, reflecting the direction of their wording relative to the rest of the items (see, e.g., Green & Citrin, 1994).

The Study

To address the directionality imbalance in the original NEP Scale and to update and broaden the scale's content, we have developed a revised NEP Scale. In keeping with the growing salience of broad "ecological" (as opposed to narrower, more specific, and less systemic "environmental") problems facing the modern world, this new and hopefully improved instrument is labeled the "New *Ecological Paradigm Scale*."

Data Collection

After being pretested with college students, the new set of NEP items was used in a 1990 mail survey of a representative sample of Washington State residents (as was the original set of items). A questionnaire covering a wide range of environmental issues was mailed out in early March of that year, and the data collection ended in early May. It proved impossible to contact 145 members of the sample of 1,300 (because of their having moved and left no forwarding addresses, being deceased, etc.), and 676 completed questionnaires were received from the remaining 1,155 potential respondents, for a completion rate of 58.5%. Given that funding allowed for only two follow-ups, rather than the recommended three (Dillman, 1978), this is a reasonably good overall response rate.

Item Construction and Modification

Besides achieving a better balance between pro- and anti-NEP statements, we also wanted to broaden the content of the scale beyond the original three facets of balance of nature, limits to growth, and antianthropocentrism. The notion of "human exemptionalism," or the idea that humans—unlike other species—are exempt from the constraints of nature (Dunlap & Catton, 1994), became prominent in the 1980s through the efforts of Julian Simon and other defenders of the DSP. In addition, the emergence of ozone depletion, climate change, and human-induced global environmental change in general suggested the importance of including items focusing on the likelihood of potentially catastrophic environmental changes or "ecocrises" besetting humankind. Consequently, we added items to tap both the exemptionalism and ecocrisis facets. Finally, we wanted to modify the outmoded sexist terminology ("mankind") present in some of the original items and decided to include an "unsure" category as a midpoint to cut down on item nonresponse.

The set of 15 items shown in Table 1 (including 6 from the original NEP Scale, 4 of which are modified very slightly) was constructed to achieve these purposes. Three items were designed to tap each of the five hypothesized facets of an ecological worldview: the reality of limits to growth (1, 6, 11), antianthropocentrism (2, 7, 12), the fragility of nature's balance (3, 8, 13), rejection of exemptionalism (4, 9, 14) and the possibility of an ecocrisis (5, 10, 15). (Item 5 was in the original NEP Scale and typically showed up in the "balance" dimension.) The eight odd-numbered items were worded so that agreement indicates a proecological view, and the seven even-numbered ones so that disagreement indicates a proecological worldview.

Table 1. Frequency Distributions and Corrected Item-Total Correlations
for New Ecological Paradigm Scale Items[a]

Do you agree or disagree[b] that:	SA[c]	MA	U	MD	SD	(N)	r_{i-t}
1. We are approaching the limit of the number of people the earth can support	27.7%	25.2%	21.0%	16.0%	10.0%	(667)	.43
2. Humans have the right to modify the natural environment to suit their needs	4.1	28.5	9.2	33.9	24.3	(663)	.35
3. When humans interfere with nature it often produces disastrous consequences	44.6	37.6	4.0	11.2	2.5	(668)	.42
4. Human ingenuity will insure that we do NOT make the earth unlivable	7.8	23.5	21.5	24.4	22.7	(664)	.38
5. Humans are severely abusing the environment	51.3	35.3	2.6	9.3	1.5	(665)	.53
6. The earth has plenty of natural resources if we just learn how to develop them	24.4	34.8	11.3	17.5	11.9	(663)	.34
7. Plants and animals have as much right as humans to exist	44.7	32.2	4.7	12.8	5.7	(665)	.46
8. The balance of nature is strong enough to cope with the impacts of modern industrial nations	1.1	7.4	11.3	30.9	49.4	(664)	.53
9. Despite our special abilities humans are still subject to the laws of nature	59.6	31.3	5.4	2.9	0.8	(664)	.33
10. The so-called "ecological crisis" facing humankind has been greatly exaggerated	3.9	17.9	13.8	25.9	38.5	(665)	.62
11. The earth is like a spaceship with very limited room and resources	38.0	36.3	7.5	13.4	4.8	(664)	.51
12. Humans were meant to rule over the rest of nature	13.5	20.4	8.2	23.9	34.0	(661)	.51
13. The balance of nature is very delicate and easily upset	45.9	32.8	5.9	14.1	1.4	(665)	.48
14. Humans will eventually learn enough about how nature works to be able to control it	3.2	20.1	24.2	27.9	24.6	(666)	.35
15. If things continue on their present course, we will soon experience a major ecological catastrophe	34.3	31.0	16.9	14.1	3.6	(667)	.62

[a]Question wording: "Listed below are statements about the relationship between humans and the environment. For each one, please indicate whether you STRONGLY AGREE, MILDLY AGREE, are UNSURE, MILDLY DISAGREE or STRONGLY DISAGREE with it."
[b]Agreement with the eight odd-numbered items and disagreement with the seven even-numbered items indicate pro-NEP responses.
[c]SA = Strongly Agree, MA = Mildly Agree, U = Unsure, MD = Mildly Disagree, and SD = Strongly Disagree.

Results

The percentage distributions for responses to each of the 15 items are shown in Table 1. As in past studies, overall there is a tendency for respondents to endorse proecological beliefs, as pluralities and often majorities (sometimes large ones) do so on every item. This is especially true for seeing the balance of nature as being threatened by human activities but is much less true for accepting the idea that there are limits to growth. There is also considerable variation in the proportions being "unsure" about the various statements, as over 20% are unsure about items 1 (on limits) and 4 and 14 (both on human exemptionalism).

Constructing a New Ecological Paradigm Scale

We were particularly interested in determining if the 15 items can legitimately be treated as measuring a single construct. A high degree of internal consistency is a necessary condition for combining a set of items into a single measure as well as an appropriate (albeit not essential) expectation for item responses constituting a reasonably coherent worldview, so we began by examining the consistency of responses to the 15 items. The last column in Table 1 shows the corrected item-total correlations for each item. All of these correlations are reasonably strong, ranging from a low of .33 to a high of .62. Not surprisingly, then, coefficient alpha is a very respectable .83. Furthermore, deletion of any of the 15 items lowers the value of alpha. Thus, the evidence from this initial survey suggests that the set of 15 items can be treated as constituting an internally consistent measuring instrument (Mueller, 1986).

Another means of assessing internal consistency is via principal-components analysis. All 15 items load heavily (from .40 to .73) on the first unrotated factor, and this factor explains 31.3% of the total variance among the items (compared to only 10% for the second factor extracted). This and the pattern of eigenvalues (4.7, 1.5, 1.2, and 1.1) suggest the presence of one major factor and thus reinforce the prior evidence concerning the internal consistency of the revised NEP Scale (Zeller & Carmines, 1980, chap. 3).

Because the dimensionality of the original NEP Scale has frequently been investigated, we employed varimax rotation to create orthogonal dimensions, and the results are shown in Table 2. When the four factors with eigenvalues greater than one are subjected to a varimax rotation, six items load most heavily on the first factor: the three ecocrisis items (5, 10, 15), two balance-of-nature items (3, 13), and one exemptionalism item (9). In addition, three other items that load most heavily on other factors have substantial cross-loadings on the first factor: one antianthropocentrism item (7), one limits-to-growth item (11), and one balance-of-nature item (8). These results suggest the first and major factor taps the balance and ecocrisis facets heavily but also incorporates the remaining three facets to

Table 2. Principal Components Analysis of NEP Items With Varimax Rotation

		Factors			
		1	2	3	4
NEP 3	(Balance)	**60**	04	07	19
NEP 5	(Eco-Crisis)	**71**	12	20	09
NEP 9	(Anti-Exempt)	**62**	20	−15	00
NEP 10	(Eco-Crisis)	**54**	**36**	27	22
NEP 13	(Balance)	**60**	00	**33**	14
NEP 15	(Eco-Crisis)	**66**	13	**35**	21
NEP 4	(Anti-Exempt)	19	**74**	05	−05
NEP 6	(Limits)	−18	**54**	**52**	11
NEP 8	(Balance)	**30**	**63**	11	21
NEP 14	(Anti-Exempt)	06	**72**	−03	18
NEP 1	(Limits)	20	−05	**76**	16
NEP 11	(Limits)	**31**	15	**75**	01
NEP 2	(Anti-Anthro)	11	10	−02	**75**
NEP 7	(Anti-Anthro)	**38**	01	10	**63**
NEP 12	(Anti-Anthro)	08	28	26	**71**
Eigenvalue		4.7	1.5	1.2	1.1
Percentage of variance		31.3	10.0	7.8	7.4

Note: Loadings of .30 and above are in bold.

some degree. The four items loading most heavily on the second factor include the remaining two exemptionalism items (4, 14), the third balance item (8), and a limits item (6), and the ecocrisis item (10) from the first factor also cross-loads heavily on this factor. Only the marginally important third and fourth factors (with eigenvalues barely above 1.0) consist of items designed to tap the same facet. The remaining two limits items (1, 11) load most heavily on the third factor, whereas the third one (6) loads almost as heavily here as it does on the second factor, and the three anthropocentrism items (2, 7, 12) load most heavily on the fourth factor.

Different researchers will have varying interpretations of the results of this analysis. Because the evidence suggests the presence of one predominant factor, and because the first three factors have items from several facets loading heavily on them, we are not inclined to create four NEP subscales measuring the four factors that emerged from the principal-components analysis and varimax rotation. Furthermore, because all 15 items load heavily on the first *unrotated* factor, have strong item-total correlations and yield an alpha of .83 when combined into a single measure, we think it is appropriate to treat them as constituting a single (revised) NEP Scale. Further, the revised NEP Scale possesses a level of internal consistency that justifies treating it as a measure of a coherent belief system or worldview. Of course, future research on differing samples is needed to confirm the appropriateness of treating the new set of 15 items as a single measure of endorsement of an ecological worldview as opposed to creating two or more dimensions of such a worldview from the NEP items. As noted earlier, differing populations will no doubt vary in the degree to which the NEP beliefs are organized into a highly consistent belief system, and in many cases it will no doubt be more appropriate to treat the NEP as multidimensional.

Predictive and Construct Validity

Because the original NEP Scale has been subjected to a good deal of testing and has been found to have considerable validity, we are not concerned about obtaining evidence on the validity of the new measure at this stage. However, the 1990 questionnaire included a number of indicators of proenvironmental (or proecological) orientation, and examining the correlations between them and the revised NEP Scale provides at least limited data on the predictive validity of the latter. Scores on the revised NEP Scale correlate significantly ($r = .61$) with scores on a 13-item measure of the perceived seriousness of world ecological problems (the higher the NEP score, the more likely the problems are seen as serious); significantly (.57) with a 4-item measure of support for proenvironment policies (the higher the NEP score, the more support for the policies); significantly (.45) with a 4-item measure of the perceived seriousness of state and community air and water pollution (the higher the NEP score, the more likely pollution is viewed as serious); and—most importantly—significantly (.31) with a 10-item measure of (self-reported) proenvironmental behaviors (more behaviors are reported by those with high NEP scores). These results, showing that the new NEP Scale is related to a wide range of ecological attitudes and behaviors, suggest that it possesses predictive validity.

Researchers have consistently found young, well-educated, and politically liberal adults to be more proenvironmental than their counterparts and have offered theoretical explanations for these findings (Jones & Dunlap, 1992). In addition, one would expect to find people with such characteristics more likely to endorse, in particular, an ecological worldview, for the reasons noted previously. Our results fit this pattern, although only political liberalism is substantially ($r = .32$) correlated with endorsement of the NEP. Age is slightly ($-.11$), albeit significantly, related to endorsement of the NEP, as is education (.10), both in the expected direction.

Other variables that are significantly ($p < .05$) correlated with scores on the revised NEP Scale include political party (.22), with Democrats having higher NEP scores; occupational sector (.13), with those employed in primary industries having lower NEP scores; income ($-.10$), which is negatively related to endorsement of the NEP; and past residence (.08), with those raised in urban areas scoring higher on the NEP. Although these correlations are quite modest, they are generally consistent with past studies of correlates of environmental concern in general and the NEP in particular. To the extent that there are sound theoretical reasons for expecting these correlations (Jones & Dunlap, 1992), as there especially are for age, education, and ideology, such findings provide some degree of construct validity for the revised NEP Scale.

Trends in Endorsement of NEP Beliefs

To our knowledge only one previous study has obtained longitudinal data on public endorsement of the NEP. Arcury and Christianson (1990) compared responses of statewide samples of Kentucky residents to the six-item version of the NEP Scale in 1984 and 1988 (the latter following a severe summer drought) and found an increase in pro-NEP responses. The increase in support for the NEP was significant, however, only in counties that had experienced water use restrictions, leading Arcury and Christianson (1990, p. 404) to conclude that "critical environmental experience can accelerate change in environmental worldview." A secondary purpose of the present study was to examine possible changes in Washington State residents' endorsement of key elements of an ecological worldview over time. Because the sample frame and data collection techniques were the same for the 1976 and 1990 surveys, we can examine trends in Washingtonians' support for the NEP over the 14-year period.

Table 3 presents the relevant data for eight items that were used in both surveys and for which the wording was either identical or changed in only minor ways. (The last two items, reflecting the ecocrisis or ecological catastrophe facet in the revised NEP Scale, were included in the 1976 questionnaire but were not incorporated into the original scale.) It should be emphasized, however, that because "unsure" was *not* used in the 1976 survey, the 1990 results have been recomputed with that response category deleted (which accounts for the difference between these figures and the results reported in Table 1). In general, there was a modest increase in Washington residents' endorsement of elements of the NEP over the 14-year period, reaching 10% on four of the eight items. The largest increase occurred on the two items that most clearly focus on the likelihood of ecological catastrophe, suggesting that the emergence of major problems such as ozone depletion and global warming have had some effect on the public. Interestingly, however, the two items dealing with ecological limits saw a decline in support, perhaps reflecting the impact of the Reagan era (which most definitely rejected the idea of limits to growth) as well as the declining salience of energy shortages.

The overall pattern of increasing endorsement of the NEP in Washington State, especially given the "ceiling effect" imposed by the relatively strong pro-NEP views expressed in 1976, provides modest support (as does the above-noted complementary trend in Kentucky) for arguments that an ecological worldview is gaining adherents (e.g., Olsen et al., 1992). Presumably, had the original data been obtained in the 1960s, or earlier, rather than in the middle of the

Table 3. Trends in Responses to Selected NEP Items by Washington Residents, 1976 and 1990

	1976	1990[a]	Change
Ecological limits			
We are approaching the limit of the number of people the earth can support. (AGREE)	73%	67%	−6%
The earth is like a spaceship with very limited room and resources.[b] (AGREE)	83	80	−3
Balance of nature			
When humans interfere with nature it often produces disastrous consequences. (AGREE)	76	86	+10
The balance of nature is very delicate and easily upset. (AGREE)	80	84	+4
Human domination			
Humans have the right to modify the natural environment to suit their needs. (DISAGREE)	62	64	+2
Ecological catastrophe			
Humans are severely abusing the environment.[c] (AGREE)	79	89	+10
The so-called "ecological crisis" facing humankind has been greatly exaggerated.[d] (DISAGREE)	57	75	+18
If things continue on their present course, we will soon experience a major ecological catastrophe.[e] (AGREE)	60	78	+18

[a]The 1990 results were computed with "Unsure" deleted, as that category was not used in 1976.
[b]The 1976 wording was "The earth is like a spaceship with *only* limited room and resources."
[c]The 1976 wording was "*Mankind is* severely abusing the environment."
[d]The 1976 wording was "The so-called 'ecological crisis' facing *mankind* has been greatly exaggerated."
[e]The 1976 wording was "If things continue on their present course, *mankind* will soon experience a major ecological catastrophe."

so-called environmental decade, the amount of change would have been far more striking (see Dunlap, 1995, for data on long-term trends in public concern for environmental quality).

Conclusion

The results reported in this article suggest that it is appropriate to treat the new set of 15 items designed to measure endorsement of an ecological worldview as constituting a single "New Ecological Paradigm Scale." The revised NEP Scale appears to be an improved measuring instrument compared to the original scale, as it (1) provides more comprehensive coverage of key facets of an ecological worldview, (2) avoids the unfortunate lack of balance in item direction of the original scale (where only four items, all dealing with anthropocentrism, were stated in an anti-NEP direction), and (3) removes the outmoded, sexist terminology in some of the original scale's items. The revised NEP Scale has slightly more internal consistency than did the original version (alpha of .83 versus .81), although this likely stems from its having three more items (as alpha tends to increase with scale length, all other things equal). Although items were selected to represent five discernible, but interrelated, facets of an ecological worldview, thus maximizing content validity, the results suggest the presence of one dominant factor in the Washington survey.

Of course, future research will be needed to address the issue of the revised NEP Scale's dimensionality, and on some samples a clearer pattern of multidimensionality will no doubt emerge and warrant creation of two or more subscales measuring distinct dimensions of the NEP. A goal for future research will be to compare the degree to which the NEP beliefs are organized coherently across different populations, including comparing patterns of multidimensionality when distinct dimensions emerge, as well as the degree to which resulting belief systems (or worldviews) influence a range of environmental attitudes, beliefs, and behaviors.

We also hope to see additional longitudinal research employing the revised NEP Scale. Although they tap primitive beliefs about humanity's relationship with the Earth, the NEP items should be responsive to personal experiences with environmental problems (as reflected by Arcury and Christianson's [1990] Kentucky study) and to information—diffused by government agencies, scientists, environmentalists and the media—concerning the growing seriousness of environmental problems. Despite the inherent complexities involved in cognitive change (see, e.g., Eagly & Kulesa, 1997), we suspect that the never-ending emergence of new scientific evidence concerning the deleterious impacts of human activities on environmental quality and the subsequent threats these pose to the welfare of humans (and other species) will generate continual pressure for adoption of a more ecological worldview. The revised NEP Scale should prove useful in tracking possible increases in endorsement of an ecological worldview, as well as in examining the effect of specific experiences and types of information in generating changes in this worldview.

References

Albrecht, D., Bultena, G., Hoiberg, E., & Nowak, P. (1982). The new environmental paradigm scale. *Journal of Environmental Education, 13*, 39–43.

Arcury, T. A. (1990). Environmental attitudes and environmental knowledge. *Human Organization, 49*, 300–304.

Arcury, T. A., & Christianson, E. H. (1990). Environmental worldview in response to environmental problems: Kentucky 1984 and 1998 compared. *Environment and Behavior, 22*, 387–407.

Arcury, T. A., Johnson, T. P., & Scollay, S. J. (1986). Ecological worldview and environmental knowledge: The "new environmental paradigm." *Journal of Environmental Education, 17*, 35–40.

Bechtel, R. B., Verdugo, V. C., & Pinheiro, J. de Q. (1999). Environmental belief systems: United States, Brazil, and Mexico. *Journal of Cross-Cultural Psychology, 30*, 122–128.

Blake, D. E., Guppy, N., & Urmetzer, P. (1997). Canadian public opinion and environmental action. *Canadian Journal of Political Science, 30*, 451–472.

Brown, L. R. (1981). *Building a sustainable society.* New York: W. W. Norton.

Caron, J. A. (1989). Environmental perspectives of Blacks: Acceptance of the "new environmental paradigm." *Journal of Environmental Education, 20*, 21–26.

Chandler, E. W., & Dreger, R. M. (1993). Anthropocentrism: Construct validity and measurement. *Journal of Social Behavior and Personality, 8*, 169–188.

Continental Group. (1982). *Toward responsible growth: Economic and environmental concern in the balance.* Stamford, CT: Author.

Cotgrove, S. (1982). *Catastrophe or cornucopia.* New York: John Wiley & Sons.

Dalton, R. J., Gontmacher, Y., Lovrich, N. P., & Pierce, J. C. (1999). Environmental attitudes and the new environmental paradigm. In R. J. Dalton, P. Garb, N. P. Lovrich, J. C. Pierce, & J. M. Whitely (Eds.), *Critical masses: Citizens, nuclear weapons production, and environmental destruction in the United States and Russia* (pp. 195–230). Cambridge, MA: MIT Press.

Dillman, D. A. (1978). *Mail and telephone surveys.* New York: Wiley Interscience.

Dunlap, R. E. (1995). Public opinion and environmental policy. In J. P. Lester (Ed.), *Environmental politics and policy* (2nd ed., pp. 63–114). Durham, NC: Duke University Press.

Dunlap, R. E. (1998). Lay perceptions of global risk: Public views of global warming in cross-national context. *International Sociology, 13*, 473–498.

Dunlap, R. E., & Catton, Jr., W. R. (1994). Toward an ecological sociology. In W. V. D'Antonio, M. Sasaki, & Y. Yonebayashi (Eds.), *Ecology, society and the quality of social life* (pp. 11–31). New Brunswick, NJ: Transaction.

Dunlap, R. E., & Van Liere, K. D. (1978). The "new environmental paradigm": A proposed measuring instrument and preliminary results. *Journal of Environmental Education, 9*, 10–19.

Dunlap, R. E., & Van Liere, K. D. (1984). Commitment to the dominant social paradigm and concern for environmental quality. *Social Science Quarterly, 65*, 1013–1028.

Eagly, A. H., & Kulesa, P. (1997). Attitudes, attitude structure, and resistance to change. In M. H. Bazerman, D. M. Messick, A. E. Tenbrunsel, & K. A. Wade (Eds.), *Environmental ethics and behavior* (pp. 122–153). San Francisco: New Lexington.

Ebreo, A., Hershey, J., & Vining, J. (1999). Reducing solid waste. Linking recycling to environmentally responsible consumerism. *Environment and Behavior, 31*, 107–135.

Edgell, M. C. R., & Nowell, D. E. (1989). The new environmental paradigm scale: Wildlife and environmental beliefs in British Columbia. *Society and Natural Resources, 2*, 285–296.

Ellis, R. J., & Thompson, F. (1997). Culture and the environment in the Pacific Northwest. *American Political Science Review, 91*, 885–897.

Furman, A. (1998). A note on environmental concern in a developing country: Results from an Istanbul survey. *Environment and Behavior, 30*, 520–534.

Gardner, G. T., & Stern, P. C. (1996). *Environmental problems and human behavior.* Boston: Allyn and Bacon.

Geller, J. M., & Lasley, P. (1985). The new environmental paradigm scale: A reexamination. *Journal of Environmental Education, 17*, 9–12.

Gooch, G. D. (1995). Environmental beliefs and attitudes in Sweden and the Baltic states. *Environment and Behavior, 27*, 513–539.

Gray, D. B. (1985). *Ecological beliefs and behaviors: Assessment and change.* Westport, CT: Greenwood Press.

Green, D. P., & Citrin, J. (1994). Measurement error and the structure of attitudes: Are positive and negative judgments opposites? *American Journal of Political Science, 38*, 256–281.

Jones, R. E., & Dunlap, R. E. (1992). The social bases of environmental concern: Have they changed over time? *Rural Sociology, 57*, 28–47.

Kempton, W., Boster, J. S., & Hartley, J. A. (1995). *Environmental values in American culture.* Cambridge, MA: MIT Press.

Lefcourt, H. M. (1996). Perspective-taking humor and authoritarianism as predictors of anthropocentrism. *Humor, 9*, 57–71.

Milbrath, L. W. (1984). *Environmentalists: Vanguard for a new society.* Albany, NY: State University of New York Press.

Mueller, D. J. (1986). *Measuring attitudes: A handbook for researchers and practitioners*. New York: Teachers College Press.

Noe, F. P., & Hammitt, W. E. (1992). Environmental attitudes and the personal relevance of management actions in a park setting. *Journal of Environmental Management, 35*, 205–216.

Noe, F. P., & Snow, R. (1989–90). Hispanic cultural influence on environmental concern. *Journal of Environmental Education, 21*, 27–34.

Noe, F. P., & Snow, R. (1990). The new environmental paradigm and further scale analysis. *Journal of Environmental Education, 21*, 20–26.

O'Connor, R. E., Bord, R. J., & Fisher, A. (1999). Risk perceptions, general environmental beliefs, and willingness to address climate change. *Risk Analysis, 19*, 461–471.

Olsen, M. E., Lodwick, D. G., & Dunlap, R. E. (1992). *Viewing the world ecologically*. Boulder, CO: Westview.

Pierce, J. C., Dalton, R. J., & Zaitsev, A. (1999). Public perceptions of environmental conditions. In R. J. Dalton, P. Garb, N. P. Lovrich, J. C. Pierce, & J. M. Whitely (Eds.), *Critical masses: Citizens, nuclear weapons production, and environmental destruction in the United States and Russia* (pp. 97–129). Cambridge, MA: MIT Press.

Pierce, J. C., Lovrich, Jr., N. P., Tsurutani, T., & Takematsu, A. (1987). Environmental belief systems among Japanese and American elites and publics. *Political Behavior, 9*, 139–159.

Pierce, J. C., Steger, M. E., Steel, B. S., & Lovrich, N. P. (1992). *Citizens, political communication and interest groups: Environmental organizations in Canada and the United States*. Westport, CT: Praeger.

Pirages, D. C., & Ehrlich, P. R. (1974). *Ark II: Social response to environmental imperatives*. San Francisco: W. H. Freeman.

Roberts, J. A., & Bacon, D. R. (1997). Exploring the subtle relationships between environmental concern and ecologically conscious consumer behavior. *Journal of Business Research, 40*, 79–89.

Rokeach, M. (1968). *Beliefs, attitudes, and values*. San Francisco: Jossey-Bass.

Schultz, P. W., & Oskamp, S. (1996). Effort as a moderator of the attitude-behavior relationship: General environmental concern and recycling. *Social Psychology Quarterly, 59*, 375–383.

Schultz, P. W., & Stone, W. F. (1994). Authoritarianism and attitudes toward the environment. *Environment and Behavior, 26*, 25–37.

Schultz, P. W., & Zelezny, L. C. (1998). Values and proenvironmental behavior: A five-country survey. *Journal of Cross Cultural Psychology, 29*, 540–558.

Scott, D., & Willits, F. K. (1994). Environmental attitudes and behavior: A Pennsylvania survey. *Environment and Behavior, 26*, 239–260.

Shetzer, L., Stackman, R. W., & Moore, L. F. (1991). Business-environment attitudes and the new environmental paradigm. *Journal of Environmental Education, 22*, 14–21.

Stern, P. C., Dietz, T., & Guagnano, G. A. (1995). The new ecological paradigm in social-psychological context. *Environment and Behavior, 27*, 723–743.

Stern, P. C., Dietz, T., Kalof, L., & Guagnano, G. A. (1995). Values, beliefs, and proenvironmental attitude formation toward emergent attitude objects. *Journal of Applied Social Psychology, 25*, 1611–1636.

Stern, P. C., Young, O. R., & Druckman, D. (1992). *Global environmental change: Understanding the human dimensions*. Washington, DC: National Academy Press.

Tarrant, M. A., & Cordell, H. K. (1997). The effect of respondent characteristics on general environmental attitude-behavior correspondence. *Environment and Behavior, 29*, 618–637.

Thompson, S. C. G., & Barton, M. A. (1994). Ecocentric and anthropocentric attitudes toward the environment. *Journal of Environmental Psychology, 14*, 149–158.

Vining, J., & Ebreo, A. (1992). Predicting behavior from global and specific environmental attitudes and changes in recycling opportunities. *Journal of Applied Social Psychology, 22*, 1580–1607.

Weigel, R. H. & Weigel, J. (1978). Environmental concern: The development of a measure. *Environment and Behavior, 10*, 3–15.

Widegren, O. (1998). The new environmental paradigm and personal norms. *Environment and Behavior, 30*, 75–100.

Zeller, R. A., & Carmines, E. G. (1980). *Measurement in the social sciences*. New York: Cambridge University Press.

RILEY E. DUNLAP is Boeing Distinguished Professor of Environmental Sociology at Washington State University and Past President of the International Sociological Association's Research Committee on Environment and Society. His research focuses on environmental attitudes, beliefs, and behaviors and the environmental movement. He has published numerous articles on these topics with his current coauthors, all of whom received their PhDs in sociolology from Washington State University.

KENT D. VAN LIERE is Chief Executive Officer of Primen, a newly formed information services company providing e-based knowledge solutions to clients interested in the deregulating energy markets. Primen is a joint venture of the Gas Research Institute and the Electric Power Research Institute. Van Liere has previously held senior positions with Hagler Bailly and HBRS, and was Associate Professor of Sociology at the University of Tennessee.

ANGELA G. MERTIG is an Assistant Professor in the Department of Sociology and the Department of Fisheries and Wildlife at Michigan State University. She is coeditor (with Dunlap) of *American Environmentalism*. Her research focuses on the environmental movement, public opinion on environmental/natural resource issues, and landscape and land use change.

ROBERT EMMET JONES is an Associate Professor in the Department of Sociology at the University of Tennessee. He teaches and conducts research on issues related to the human dimensions of environmental change and ecosystem management. He has published articles on these topics in journals such as *Rural Sociology*, *Social Science Quarterly*, and the *Journal of Environmental Systems*.

Journal of Social Issues, Vol. 56, No. 3, 2000, pp. 443–457

Elaborating on Gender Differences in Environmentalism

Lynnette C. Zelezny*

California State University, Fresno

Poh-Pheng Chua

Pennsylvania State University

Christina Aldrich

Claremont Graduate University

A review of recent research (1988 to 1998) on gender differences in environmental attitudes and behaviors found that, contrary to past inconsistencies, a clearer picture has emerged: Women report stronger environmental attitudes and behaviors than men. Additional evidence of gender differences in environmental attitudes and behaviors was also supported across age (Study 1) and across 14 countries (Study 2). As a single variable, the effect of gender on proenvironmental behavior was consistently stronger than on environmental attitudes. Explanations for gender differences in environmentalism were examined in Study 3. It was found that compared to males, females had higher levels of socialization to be other oriented and socially responsible. Implications for theory, social action, and policy are discussed.

One of the ways psychologists can promote environmentalism is to understand the relationship between demographic variables and environmental attitudes and behaviors and the implications these human-environment relationships may have on theory, social action, and policy. Numerous studies have examined the relationship between demographic variables (e.g., age, education, ethnicity, socioeconomic status) and environmental attitudes and behaviors. Research on

*Correspondence concerning this article should be addressed to Lynnette Zelezny at Department of Psychology, California State University, 5310 North Campus Drive, M/S PH 11, Fresno, CA [e-mail: lynnette@csufresno.edu].

environmentalism and gender has been somewhat limited, however, and "surprisingly little has been done to examine the . . . environmental activity of women and factors related to it" (Mohai, 1992, p. 2).

Two recognized reviews of gender difference in environmental attitudes and behaviors conducted more than a decade ago (Hines, Hungerford, & Tomera, 1986–87; Van Liere & Dunlap, 1980) concluded that research on the relationship between these variables is meager and inconsistent. In concurrence, Mohai (1992) stated that "no firm conclusions can be drawn about the effects of gender on concern about general environmental issues, and more analysis and explanation clearly needs to be done in this area" (p. 2).

Review of Literature on Gender Differences and Environmentalism

To elaborate on these past findings, we surveyed a decade of research, from 1988 to 1998, on gender and environmental attitudes and behaviors since Hines et al.'s (1986–87) meta-analytic review. We found that numerous studies ($n = 32$) had recently been added to the literature. We focused primarily on published studies that had measured environmental attitudes and/or behaviors. Most studies that examined gender and environmental attitudes typically measured environmental attitudes using a single item (e.g., Blocker & Eckberg, 1989, 1997; Mohai, 1992); however, a few ($n = 6$) included the New Environmental Paradigm (NEP) Scale (Dunlap & Van Liere, 1978; Dunlap, Van Liere, Mertig, & Jones, 2000). Our review focused on the six studies that used the NEP to measure environmental attitudes (i.e., Arcury, 1990; Arcury & Christianson, 1990, 1993; Blaikie, 1992; Maineri, Barnett, Valdero, Unipan, & Oskamp, 1997; Widegren, 1998). Descriptive analyses showed that four of the six studies found that females expressed significantly greater (NEP) environmental concern than males (Arcury, 1990; Arcury & Christianson, 1990; Blaikie, 1992; Maineri, Barnett, Valdero, Unipan, & Oskamp, 1997), whereas two of the of six studies found no significant difference between males and females on (NEP) environmental concern (Arcury & Christianson, 1993; Widegren, 1988); no study found that males had significantly greater NEP environmental concern than women. Using meta-analytic techniques across the six studies mentioned above, the calculated effect of gender on NEP environmental attitudes was $r = .07$. Thus, it was concluded that the majority of studies from 1988 to 1998 found that women reported significantly more general environmental concern than men, although the effect of gender on NEP environmental attitudes was small.

In terms of gender and environmental behavior we identified 13 studies published since Hines et al.'s 1987 review (i.e., Arcury & Christianson, 1993; Arp & Howell, 1995; Baldassare & Katz, 1992; Blocker & Eckberg, 1997; Maineri, Barnett, Valdero, Unipan, & Oskamp, 1997; Mohai, 1992; Roberts, 1993; Schahn & Holzer, 1990; Steel, 1996; Stern, Dietz, & Kalof, 1993; Stern, Dietz, Kalof, &

Guagnano, 1995; Widegren, 1998; Wolkomir, Futreal, Woodrum, & Hoban, 1997). Of these 13 studies, 9 found that women reported significantly more participation in proenvironmental behaviors than men (Baldassare & Katz, 1992; Maineri, Barnett, Valdero, Unipan, & Oskamp, 1997; Roberts, 1993; Schahn & Holzer, 1990; Steel, 1996; Stern, Dietz, & Kalof, 1995; Stern, Dietz, Kalof, & Guagnano, 1995; Widegren, 1998; Wolkomir, Futreal, Woodrum, & Hoban, 1997); 3 of the 13 studies found no significant difference between males and females on proenvironmental behavior (Arcury & Christianson, 1993; Arp & Howell, 1995; Blocker & Eckberg, 1997); and 1 of the 13 studies found that males reported significantly greater participation in proenvironmental behavior (Mohai, 1992). Thus, the majority of studies in the last decade found that, compared to men, women reported greater participation in proenvironmental behavior/activism. Meta-analytic techniques were also utilized here to clarify the relationship between gender and proenvironmental behaviors. Across these studies it was found that the effect of gender on proenvironmental behavior was $r = .10$.

Why Are Females More Environmental?

A variety of theories have been used to explain gender differences in environmentalism. One widely used approach is based on gender roles and socialization (Eagly, 1987; Howard & Hollander, 1996; Miller, 1993; Unger & Crawford, 1996; Wilkinson & Kitzinger, 1996). Socialization theory posits that behavior is predicted by the process of socialization, whereby individuals are shaped by gender expectations within the context of cultural norms. Females across cultures are socialized to be more expressive, to have a stronger "ethic of care," and to be more interdependent, compassionate, nurturing, cooperative, and helpful in caregiving roles (Beutel & Marini, 1995; Chodorow, 1974; Eagly, 1987; Gilligan, 1982). On the other hand, males are socialized to be more independent and competitive (Chodorow, 1974; Gilligan, 1982; Keller, 1985).

Theoretically, gender differences in environmentalism imply links between socialization and values (Stern, Dietz, & Kalof, 1993). As guiding principles, values (Rokeach, 1973) predict attitudes and behaviors (Olson & Zanna, 1994); therefore, because females, compared to males, are socialized to value the needs of others, women exhibit more helping behavior and altruism (Gilligan, 1982).

Individuals who help possess an "other" value orientation according to Schwartz's (1968, 1977) norm activation model. This model suggests that helping behavior is most likely to occur when individuals are aware of harmful consequences (awareness of consequences—AC) and of their actions and feel responsible for these consequences (ascribed responsibility—AR). Schwartz's norm activation model has been applied to the environmental domain to explain proenvironmental behavior (Schultz & Zelezny, 1998; Black, Stern, & Elworth, 1985; Heberlein & Black, 1976; Hopper & Nielson, 1991; Stern, Dietz, & Black,

1986; Van Liere & Dunlap, 1978). Because proenvironmental behavior is a special kind of helping, however, other value orientations may underlie helping the environment (i.e., ecological value orientation; Stern, Dietz, & Kalof, 1993). Ecological value orientation is defined as the expression of motivational concern for environmental issues based upon an individual's conception of humanity's relationship to the environment (Thompson & Barton, 1994). Stern and Dietz (1994) proposed a tripartite classification of ecological value orientations: concern for self, concern for other human beings, and concern for the biosphere. Thus, individuals may act proenvironmentally but they may have different values. For example, individuals may protest agricultural pollution (e.g., spraying pesticides) because it affects the air they breathe (i.e., egocentric orientation) or because it affects the air their children breathe (i.e., anthropocentric orientation) or because it affects the ecosystem (i.e., ecocentric orientation). Stern, Dietz, and Kalof (1993) proposed that individuals may have environmental attitudes that reflect a combination of these three value orientations. They found that women have stronger beliefs about the harmful consequences of poor environmental conditions for others, the biosphere, and self and that these beliefs predicted more proenvironmental behavior (Stern, Dietz, & Kalof, 1993).

For this article, we were particularly interested in ecocentrism, which is a fundamental belief in the inherent value of nature, the biosphere, and all living things. By definition ecocentrism suggests an extended "other" orientation, which, according to gender socialization theory, is characteristic of feminine socialization. Numerous studies have found that gender socialization significantly influences individual behavior very early in life, and these findings have been supported across cultures (Block, 1973; Williams & Best, 1990). In a recent study on gender and the environmental risk concerns, however, Davidson and Freudenberg (1996) suggested that gender differences in environmentalism are not universal. Therefore, we wondered whether gender differences in environmentalism could be found in children and across countries. The two studies reported below address these issues.

First, in Study 1, we examined gender differences in environmental attitudes and behaviors among primary and secondary school children. Based on socialization theory, we predicted that, compared to males, female students in primary and secondary schools would report significantly stronger environmental attitudes and greater participation in proenvironmental behaviors. Study 2 examined gender differences in environmental attitudes and behaviors across 14 countries to test Davidson and Freudenberg's (1996) conclusions that gender differences in environmentalism are not universal.

Based on the aforementioned socialization framework and past research on other versus self-orientation, we derived three hypotheses. First, we predicted that females across countries would report significantly stronger environmental attitudes as measured by the NEP. Secondly, we predicted that females across countries would report stronger value-based ecocentric environmental attitudes than males. Third, we predicted that females across countries would report greater participation in proenvironmental behaviors than males.

Study 1: Gender Differences in Environmentalism Among Children

A stratified sample of primary and secondary school students (N_{1994} = 584; N_{1995} = 709) from diverse socioeconomic strata in California were systematically surveyed over a 2-year period.

A 35-item questionnaire was designed to assess students' (1) general environmental attitudes (measured by separate items and an adapted NEP Scale; Dunlap & Van Liere, 1978), (2) self-reported knowledge about the environment, (3) feelings of personal responsibility for improving the environment, (4) specific environmental attitudes, (5) specific recycling attitudes, (6) interest and intention to participate in school recycling, (7) participation in school recycling after the initiation of district-wide school recycling (1995 survey only), and (8) demographic characteristics. The questionnaire included both closed-ended questions (n = 32) and open-ended questions (n = 3). The closed-ended questions, designed by the current authors, included 6-point Likert-type response scales (e.g., in response to question 6, "How would you rate your overall personal responsibility to improve the environment?" the response choices were 1—*Extremely responsible*, 2—*Very responsible*, 3—*Somewhat responsible*, 4—*Slightly responsible*, 5—*Not at all responsible*, and 6—*I don't know*).

The survey was pilot tested twice on elementary school students to revise the wording of the survey items so that they were appropriate for children. In the pilot testing, half of the items from the full 12-item scale of the NEP (Dunlap & Van Liere, 1978) were not understood by younger elementary students, and those items were not included in the final questionnaire. A shortened six-item NEP Scale assessed students' general environmental concern using the following items: (1) The balance of nature is very delicate and easily upset; (2) People must live in harmony with nature in order to survive; (3) Pollution is not personally affecting my life; (4) Courses focusing on conservation of natural resources should be taught in the public schools; (5) Although there is contamination of our lakes, streams, and air, nature will soon return them to normal; and (6) Because government rules are so effective, it is not likely that pollution will become too bad. The shortened NEP included Likert response choices as follows: (1) *Strongly agree*, (2) *Agree*, (3) *No opinion*, (4) *Disagree*, and (5) *Strongly disagree*. It was also determined, because of the reading level required to adequately comprehend this questionnaire, that the minimum grade level of students in the study would be fifth grade.

The reliabilities of the student environmental questionnaire were assessed separately using the 1994 and 1995 data. The reliability coefficients ranged from alpha .87 to .88 on all self-designed environmental questions and the shortened

NEP and from alpha .71 to .72 on the shortened NEP scale, across the 2 years. The reliability coefficients for this questionnaire remained relatively stable across ages, schools, ethnicities, gender, and socioeconomic status; the most consistent responses for these particular environmental items, however, came from Anglo high school students of middle or high socioeconomic status.

The relationships between students' demographic characteristics (e.g., age and gender) and their general environmental concern, specific environmental attitudes, and participation in school recycling were analyzed. The environmental attitudes and behaviors of girls and boys were compared by year, as shown in Table 1. In addition, the effect of gender on NEP attitudes and proenvironmental behavior was assessed using effect size (r) calculations (see Table 2). Finally, gender differences in concern about specific environmental issues—air pollution, animal extinction, cutting down trees, wasting energy, water pollution, and trash in the environment—were qualitatively compared by year.

In 1994, with regard to general attitudes, girls reported significantly stronger overall concern about the environment than boys. Further, girls reported significantly stronger general environmental concerns than boys on the NEP and more personal responsibility to improve the environment. In terms of specific attitudes, girls expressed greater proenvironmental attitudes than did boys on concern about trash, interest in recycling, and interest in school recycling. Finally, girls reported stronger intentions to participate in school recycling than did boys (see Table 1).

In 1995, the pattern was identical. Girls reported stronger overall concern for the environment, general NEP environmental concern, and personal responsibility

Table 1. Comparison of Girls and Boys on Environmental Attitudes and Behavior, by Year

DV	Girls $n_{1994} = 303$ $n_{1995} = 353$	Boys $n_{1994} = 260$ $n_{1995} = 337$	F ratio
Q3. Self-rated environmental concern	3.49 **3.36**	3.27 **3.13**	7.00** **8.81****
Q6. Self-rated personal responsibility to improve the environment	3.34 **3.25**	3.07 **2.91**	8.63** **17.85****
Q12. Concern about trash	4.11 **4.07**	3.81 **3.76**	12.39** **14.22****
Q14. Interest in recycling	3.94 **3.85**	3.72 **3.69**	8.11** **4.47****
Q21. Interest in school recycling	4.21 **3.90**	3.95 **3.61**	11.04** **14.60****
Q22. Participation in school recycling	3.88 **3.31**	3.48 **2.91**	17.25** **15.92****
NEP Scale	22.93 **22.68**	21.94 **21.52**	8.95** **13.46****

Note: 1994 means are the upper values in each cell. 1995 means are the lower values in each cell and are in boldface.
* $p < .05$. ** $p < .01$.

Table 2. Comparison of Gender Effects (Female) on NEP Environmental Attitudes and Proenvironmental Behaviors, by Study

Study	Variable	Effect size (r)
Review of studies 1988–1999	NEP environmental attitudes	$r = .07$
Study 1 Schoolchildren 1994 Schoolchildren 1995	NEP environmental attitudes	$r = .13$ $r = .14$
Study 2 University students in 14 countries	NEP environmental attitudes	$r = .04$
Study 2 University students in 14 countries	Ecocentric environmental attitudes	$r = .10$
Review of studies from 1988–1999	Proenvironmental behaviors	$r = .10$
Study 1 Schoolchildren 1994 Schoolchildren 1995	Proenvironmental behaviors	$r = .18$ $r = .15$
Study 2 University students in 14 countries	Proenvironmental behaviors	$r = .09$

for improving the environment than boys. Further, girls reported stronger concern about trash, interest in recycling, and interest in school recycling. Finally, girls reported significantly more participation in school recycling (see Table 1).

The effect sizes of gender on environmental variables were calculated using the results from Study 1. The effect of gender on environmental attitudes as measured by the NEP ranged from $r = .13$ to $r = .14$ (see Table 2). The effect of gender on proenvironmental behaviors ranged from $r = .18$ to $r = .15$ (see Table 2).

Qualitatively, with regard to specific environmental issues, girls reported in both 1994 and 1995 that the issue that they cared the most about was animal extinction. Boys, however, reported in 1994 that their top concern was animal extinction, whereas in 1995, they reported that they were most concerned about water pollution. On the other hand, girls and boys consistently reported, across both years, that they were least concerned about wasting energy.

In summary, compared to boys, girls reported stronger proenvironmental responses on all environmental variables in this study, and this pattern was consistent across 2 years (see Table 1).

The findings in Study 1 strongly suggest that environmentalism does not begin in adulthood, thus debunking the argument that gender differences in environmentalism arise with motherhood and protecting children from environmental threats (Hamilton, 1985a; Levine, 1982). Study 1's findings are consistent with the adult studies that were reviewed earlier in this article. Females, regardless of age (i.e., youth or adult) reported more concern for the environment and proenvironmental behaviors than males. Also, identical patterns emerged with respect to effect sizes. In both adults and youth, the effect of gender (female) was stronger on proenvironmental behaviors than NEP environmental concerns (see Table 2).

We are careful to note the limitations of our investigation. We did not directly compare the environmental attitudes and behaviors of children and adults, nor did we examine environmental attitudes and behaviors longitudinally within individuals. Future research is needed to understand the development of environmentalism, the stability of environmentalism across the life span, the interaction between age and gender on environmental attitudes and behaviors, and how environmentalism is related to cognitive, moral, and social development. Finally, research on gender differences and environmentalism is needed to address the generalizability and the universality of these findings.

Hence, in response to the recommendations presented in Study 1, we report data from a study to examine the universality of gender differences and environmentalism across countries.

Study 2: Gender Differences in Environmentalism Across 14 Countries

English- and Spanish-speaking students ($n = 2,160$) from Europe, Latin America, and the United States were contacted via professors and administrators from universities throughout the world. All students were undergraduates participating in a social or behavioral studies course in the following 14 countries: Argentina ($n = 54$), Canada ($n = 96$), Colombia ($n = 149$), Costa Rica ($n = 213$), the Dominican Republic ($n = 121$), Ecuador ($n = 201$), El Salvador ($n = 194$), Mexico ($n = 65$), Panama ($n = 100$), Paraguay ($n = 200$), Peru ($n = 224$), Spain ($n = 104$), the United States ($n = 245$), and Venezuela ($n = 194$). Males and females were represented from each country; our total sample included 781 males (36%) and 1,379 females (64%). In our sample, 16% ($n = 341$) were from countries whose native language is English (Canada and the United States), and 84% ($n = 1,819$) were from countries whose native language is Spanish (Argentina, Colombia, Costa Rica, the Dominican Republic, Ecuador, El Salvador, Mexico, Panama, Peru, Paraguay, Spain, and Venezuela). The average student age for the total sample was 24.72. Relative socioeconomic status of students in this sample was assessed based on a single question that asked "Relative to the people in your country, would you say that your family is 1 (lower class) to 10 (upper class)?" The average self-reported socioeconomic status rating for our sample was 4.72. See Schultz and Zelezny (1999) for more detail on the sample.

A four-page questionnaire was designed to assess students' demographic characteristics, general environmental attitudes, value-based environmental attitudes, and proenvironmental behaviors. General environmental attitudes were measured using the revised 15-item NEP Scale (Dunlap et al., 2000), which includes a 5-point Likert response scale that ranges from 1 (*strongly agree*) to 5 (*strongly disagree*). Value-based environmental attitudes were measured using 14 items from Thompson and Barton's (1994) scale. Of particular interest here were value-based ecocentric environmental attitudes, which reflect the belief that the environment should be preserved because of the intrinsic value of the biosphere and of all living things. Specifically, value-based ecocentric environmental attitudes were measured by seven items (items 2, 5, 7, 26, 30, 32, and 33 from Thompson & Barton's scale) using a 5-point Likert response scale that ranged from 1 (*strongly agree*) to 5 (*strongly disagree*). Environmental behaviors were measured by 12 questions designed by the current authors that asked about students' past participation (i.e., daily, weekly, and monthly) in proenvironmental behaviors (e.g., political activism, recycling, energy conservation, water conservation, purchasing environmentally safe products, and using public transportation). These behavioral items included a 5-point Likert-type response scale. Finally, demographic characteristics (e.g., gender, age, socioeconomic status, education, and strength of religious beliefs) were assessed using a combination of response scales (categorical choice [gender, age, education], religious beliefs).

The questionnaire was pilot tested, translated from English to Spanish, and translated back from Spanish to English to maximize measurement validity. Questionnaires were group-administered in classes. Students participated voluntarily. Neither the students nor the contacts were monetarily compensated for participation, although in some cases students were given extra credit in courses.

We used descriptive and inferential statistical techniques to test our hypotheses. We predicted that, compared to males, females across countries would report significantly stronger environmental attitudes as measured by the NEP Scale. Our prediction was partly supported.

Using descriptive analysis, we found that interesting patterns in NEP environmental attitudes emerged. Females reported higher NEP environmental attitudes than males in 10 of the 14 countries (Argentina, Canada, Costa Rica, the Dominican Republic, Mexico, Panama, Paraguay, Peru, Spain, and the United States); males had higher NEP environmental attitudes than females in 3 of the 14 countries (Colombia, Ecuador, and El Salvador); and males and females did not differ on NEP environmental attitudes in 1 of the 14 countries (Venezuela). We found that females overall in our 14-country sample reported significantly stronger NEP environmental attitudes than males, $F(1, 1870) = 4.24$, $p < .001$. When we analyzed gender differences by country, however, which substantially reduced our statistical power, we found significant gender differences in NEP environmental attitudes only in the United States.

We also predicted that females would report significantly higher levels of value-based ecocentric environmental attitudes than males. This prediction was partly supported. Compared to males, females reported stronger ecocentric environmental attitudes in 12 of the 14 countries (Argentina, Canada, Colombia, Costa

Rica, El Salvador, Mexico, Panama, Paraguay, Peru, Spain, the United States, and Venezuela); and males reported higher ecocentric environmental attitudes than females in 2 of the 14 countries (the Dominican Republic and Ecuador). Overall we found that females in our 14-country sample did report greater ecocentric environmental attitudes than males, $F(1, 2042) = 20.43$, $p < .001$. However, significant gender differences in ecocentric environmental attitudes within countries were found only in Argentina and Panama.

Finally, we predicted that females would report greater participation in proenvironmental behaviors than males, which was partly supported. Females reported greater participation in proenvironmental behaviors than males in 11 of the 14 countries (Argentina, Canada, Costa Rica, Ecuador, El Salvador, Mexico, Paraguay, Peru, Spain, the United States, and Venezuela); and males reported greater participation in proenvironmental behavior than females in 3 of the 14 countries (the Dominican Republic, Colombia, and Panama). Overall analyses revealed that females in our 14-country sample did report greater participation in proenvironmental behaviors than males, $F(1, 1870) = 14.64$, $p < .001$; within countries, however, significant gender differences in proenvironmental behavior were found only in Paraguay and Venezuela.

The effect sizes of gender on environmental variables were calculated using the results from Study 2. The effect of gender on NEP environmental attitudes was $r = .04$, on value-based ecocentric environmental attitudes was $r = .10$, and on proenvironmental behaviors was $r = .09$ (see Table 2).

As a group, females across 14 countries did, in fact, report significantly stronger NEP environmental attitudes, stronger value-based ecocentric environmental attitudes, and greater participation in proenvironmental behaviors, although gender differences in environmental attitudes and behaviors within countries were less convincing. When significant gender differences in environmental attitudes and behaviors were found within countries, however, females were consistently more proenvironmental than men. Moreover, the descriptive patterns among countries were also notable; among the range of average scores, females consistently reported higher ratings than males on all variables, including proenvironmental behaviors (e.g., political activism).

These findings attenuate Davidson and Freudenberg's (1996) claim that gender differences in environmentalism are not universal. Gender differences in environmentalism were found across 14 countries, not just the United States, which leads us to question the role of gender socialization in environmentalism. We are careful, however, to note the limitations of our investigation. Our sample in Study 2 was not representative in many ways. All of the individuals in this multicountry study were university students; therefore, sampling bias and generalizability were a concern. We recognize that female university students, especially those from underdeveloped countries, may have been more nontraditonal in their gender attitudes and behaviors, which would result in weak gender effects, as we found in

Study 2. Further, we did not find gender differences in environmental attitudes and behaviors within countries, which we attributed partly to lack of statistical power within countries.

In our view, Studies 1 and 2 advance past studies on gender and environmentalism because we examined a broader spectrum of males and females, across ages and countries, on both environmental attitudes and behaviors, and we found very consistent patterns. Females reported stronger environmental attitudes and behaviors than men across ages and countries. Other interesting findings also emerged. We found that as a single variable, the effect of gender on environmental attitudes and behaviors, was strongest among young people; in addition, the effect of gender on proenvironmental behavior was consistently stronger than on environmental attitudes (see Table 2).

Given that these studies provide new evidence that describes gender differences in environmentalism, a more difficult question remains: How can we empirically explain these differences? Using gender socialization theory, which was discussed earlier in this article, we developed Study 3 to examine the role of socialization on gender differences and environmentalism. We predicted that, compared to males, females would have a stronger "extended other" orientation and a stronger social "ethic of care" to take responsibility for alleviating problems in the world, as theorized by Gilligan (1982) and others.

Study 3: Explaining Gender Differences and Environmentalism

University students ($N = 119$; 79 females and 40 males) volunteered to participate in this survey on environmental attitudes, gender orientation, socialized other orientation, and social responsibility. The average age of participants was 20.54.

A questionnaire was designed to measure general environmental attitudes, using the NEP Scale (Dunlap & Van Liere, 1978); feminine and masculine orientation, using the California Psychological Inventory (CPI) Femininity and Masculinity scale; the ability to take the role of a conceptualized other, using the CPI Socialization scale; and ethic of care to take responsibility for ameliorating social problems, using the Minnesota Multiphasic Personality Inventory (MMPI) Social Responsibility subscale. In addition, demographic (i.e., gender, age, perceived socioeconomic status, political affiliation) information was assessed. Questionnaires were group administered.

Using t-test analyses, we again found that females ($M = 36.89$, $SD = 4.29$) reported significantly more NEP concern for the environment than males ($M = 34.33$, $SD = 5.90$), $t(117) = -2.702$, $p = .008$. Post hoc analyses using estimate of omega-squared showed that 5% of the variance in general NEP environmental concern was accounted for by gender. These findings were consistent with the previously cited findings in Studies 1 and 2 (see Table 2).

Hypothesis 1, which predicted gender differences in affinity to take the role of a conceptualized other, was supported. Compared to males ($M = 30.14$, $SD = 6.63$), females ($M = 33.66$, $SD = 5.18$) were more able to take the role of a conceptualized other as measured by the CPI Socialization scale, $t(109) = -3.034$, $p = .003$. Estimate of omega-squared showed that 6.88% of the variance in one's ability to take the role of a conceptualized other was explained by gender.

Hypothesis 2, which predicted gender differences in social responsibility, was also supported. Compared to males ($M = 94.68$, $SD = 13.94$), females ($M = 102.75$, $SD = 10.72$) reported stronger levels of social responsibility, as measured by the MMPI Social Responsibility subscale, $t(112) = 3.417$, $p = .001$. Gender accounted for 8.56% of the variance in social responsibility, using estimate of omega-squared post hoc analyses.

In addition, post hoc analyses showed that CPI Femininity was positively correlated with general NEP environmental attitudes ($r = .249$, $p = .008$); however, CPI Masculinity was negatively correlated with NEP concern ($r = -.273$, $p = .004$). Notably, the effect of feminine orientation, as shown in Study 3, was stronger than the effect of gender, as shown in Studies 1 and 2, on NEP environmental attitudes (see Table 2). Future research is recommended to examine the role of gender orientation in environmentalism.

Summary and Conclusions

In summary, Study 3's findings implicitly support gender socialization as an explanation for gender differences in environmentalism. Study 3 corroborated the findings in Studies 1 and 2 and other gender socialization research. Like Studies 1 and 2, Study 3 found gender differences in NEP environmental concern. In addition, this study found that, compared to males, females had a stronger ability to take the role of a conceptualized other (i.e., other orientation), which was also supported by Gough (1960, 1994), and stronger levels of social responsibility, as reported by Borden and Francis (1978).

In addition, the findings in Studies 1 and 2 also indirectly support gender socialization as an explanation for gender differences in environmentalism. Females across 14 countries in Study 2 reported stronger ecocentrism (concern for nature, the biosphere, and all living things) than males, which in our view represents an "other orientation." And in Study 1, female youth, compared to male youth, reported stronger personal responsibility for improving the environment, and these findings were consistent across 2 years.

Of course, there may be other explanations for gender differences in environmental attitudes and behaviors. Our studies relied on self-reports, therefore, we wondered whether gender differences in socially desirable responding might explain these findings. However, a recent paper by Zelezny and Yelverton (2000) found that social desirability was not significantly related to gender, NEP environmental attitudes, or intentions to act proenvironmentally.

We also recognize the potential for generation effects; that is, the relationship between gender and environmentalism may have changed over the past years. However, if this were so, it would suggest that gender differences in environmentalism are likely due to socialization rather than inherent biological differences, as described by ecofeminists and others. Meta-analytic research that compares past and present studies on gender and environmental attitudes and behaviors is needed to clarify the emergence of change over time.

Finally, the findings in this article raise important new questions for future research: (1) What are the mediating links between gender, socialization, and social responsibility and environmental concern? (2) How can the impact of gender socialization on environmentalism be directly examined? (3) What is the role of parenting on gender socialization and environmentalism? and (4) Are psychologically androgynous individuals, who attain higher levels of moral development, more proenvironmental?

In conclusion, we assert that a clearer picture has emerged with regard to gender and environmentalism. Contrary to past research findings, females are not passive, indifferent, or unconcerned about the environment. Moreover, we believe that these findings have implications for theory, social action, and policy. Specifically, we project that future models of environmentalism will include gender as a relevant predictor of environmentalism and that collectively females will be influential in future environmental activism, policy development, and political leadership. Surely, environmental improvement will require the collective, conscientious effort of men, women, and children from all nations of the world. Our studies simply support the idea that females will play an active and positive role in this progress.

References

Arcury, T. A. (1990). Environmental attitudes and environmental knowledge. *Human Organization*, *49*, 300–304.

Arcury, T. A., & Christianson, E. H. (1990). Environmental worldview in response to environmental problems. *Environment and Behavior*, *22*, 387–407.

Arcury, T. A., & Christianson, E. H. (1993). Rural and urban differences in environmental knowledge and action. *Journal of Environmental Education*, *25*, 19–25.

Arp, W., & Howell, C. (1995). Black environmentalism and gender differences: An ethic of care? *The Western Journal of Black Studies*, *19*, 300–305.

Baldassare, M., & Katz, C. (1992). The personal threat of environmental problems as a predictor of environmental practices. *Environment and Behavior*, *24*, 602–617.

Beutel, A., & Marini, M. (1995). Gender and values. *American Sociological Review*, *60*, 436–448.

Black, J., Stern, P., & Elworth, J. (1985). Personal and contextual influences on household energy adaptations. *Journal of Applied Psychology*, *70*, 3–21.

Blaikie, N. (1992). The nature and origins of ecological worldviews: An Australian study. *Social Science Quarterly*, *73*, 144–165.

Block, J. H. (1973). Conceptions of sex roles: Some cross-cultural and longitudinal perspectives. *American Psychologist*, *28*, 512–526.

Blocker, T., & Eckberg, D. (1989). Environmental issues as women's issues: General concerns and local hazards. *Social Science Quarterly, 70,* 586–593.

Blocker, T., & Eckberg, D. (1997). Gender and environmentalism: Results from the 1993 General Social Survey. *Social Science Quarterly, 78,* 841–858.

Borden, R. J., & Francis, J. L. (1978). Who cares about ecology? Personality and sex differences in environmental concern. *Journal of Personality, 46,* 190–203.

Chodorow, N. (1974). Family structure and feminine perspective. In M. Rosaldo & L. Lamphere (Eds.), *Women in culture and society* (pp. 41–48). Stanford, CA: Stanford University Press.

Davidson, D., & Freudenberg, W. (1996). Gender and environmental risk concerns: A review and analysis of available research. *Environment and Behavior, 28,* 302–339.

Dunlap, R., & Van Liere, K. (1978). The new environmental paradigm. *Journal of Environmental Education, 9,* 10–19.

Dunlap, R., Van Liere, K., Mertig, A., & Jones, R. E. (2000). Measuring endorsement of the New Ecological Paradigm: A revised NEP scale. *Journal of Social Issues, 56*(3), 425–442.

Eagly, A. (1987). *Sex differences in social behavior: A social role interpretation.* Hillsdale, N.J.: Erlbaum.

Gilligan, C. (1982). *In a different voice.* Cambridge, MA: Harvard University Press.

Gough, H. G. (1960). Theroy and measurement of socialization. *Journal of Consulting Psychology, 24,* 23–30.

Gough, H. G. (1994). Theory, development, and interpretation of the CPI socialization scale. *Psychological Reports, 75,* 651–700.

Hamilton, L. (1985a). Concerns about toxic wastes: Three demographic predictors. *Sociological Perspectives, 28,* 463–486.

Hamilton, L. (1985b). Who cares about water pollution? Opinions in a small-town crisis. *Sociological Inquiry, 55,* 170–181.

Heberlein, T., & Black, J. (1976). Attitudinal specificity and the prediction of behavior in a field setting. *Journal of Personality and Social Psychology, 33,* 474–479.

Hines, J., Hungerford, H., & Tomera, A. (1986–87). Analysis and synthesis of research on responsible environmental behavior. *Journal of Environmental Education, 18,* 1–8.

Hopper, J., & Nielson, J. (1991). Recycling as altruistic behavior: Normative and behavioral strategies to expand participation in a community recycling program. *Environment and Behavior, 23,* 195–220.

Howard, J. A., & Hollander, J. A. (1996). *Gendered situations, gendered selves.* Thousand Oaks, CA: Sage.

Keller, E. (1985). *Reflections on gender and science.* New Haven, CT: Yale University Press.

Levine, A. (1982). *Love Canal.* Lexington, MA: Lexington Books.

Maineri, T., Barnett, E., Valdero, T., Unipan, J., & Oskamp, S. (1992). Green buying: The influence of environmental concern on consumer buying. *Journal of Social Psychology, 137,* 189–204.

Miller, B. D. (Ed.). (1993). *Sex and gender hierarchies.* New York: Cambridge University Press.

Mohai, P. (1992). Men, women, and the environment. *Society and Natural Resources, 5,* 1–19.

Nelkins, D. (1981). Nuclear power as a feminist issue. *Environment, 23,* 14–39.

Olson, J., & Zanna, M. (1994). Attitudes and attitude change. *Annual Review of Psychology, 44,* 117–154.

Roberts, J. A. (1993). Sex differences in socially responsible consumers' behavior. *Psychological Reports, 73,* 139–148.

Rokeach, M. (1973). *The nature of human values.* New York: Free Press.

Schahn, J., & Holzer, E. (1990). Studies of individual environmental concern: The role of knowledge, gender, and background variables. *Environment and Behavior, 22,* 767–786.

Schultz, P. W., & Zelezny, L. C. (1998). Values and proenvironmental behavior: A five-country survey. *Journal of Cross-Cultural Psychology, 29,* 540–558.

Schultz, P. W., & Zelezny, L. C. (1999). Vaules as predictors of environmental attitudes: Evidence for consistency across cultures. *Journal of Environmental Psychology, 19,* 255–265.

Schwartz, S. H. (1968). Words, deeds, and the perception of consequences and responsibility in action situations. *Journal of Personality and Social Psychology, 10,* 232–242.

Schwartz, S. H. (1977). Normative influences in altruism. In L. Berkowitz (Ed.), *Advances in experimental social psychology* (Vol. 10, pp. 221–279). New York: Academic Press.

Steel, B. S. (1996). Thinking globally and acting locally? Environmental attitudes, behavior, and activism. *Journal of Environmental Management, 47*, 27–36.

Stern, P., & Dietz, T. (1994). The value basis of environmental concern. *Journal of Social Issues, 26*, 1–20.

Stern, P., Dietz, T., & Black, J. (1986). Support for environmental protection: The role of moral norms. *Population and Environment, 8*, 204–222.

Stern, P., Dietz, T., & Kalof, L. (1993). Value orientations, gender, and environmental concern. *Environment and Behavior, 25*, 322–348.

Stern, P., Dietz, T., Kalof, L., & Guagnano, G. (1995). Values, beliefs, and proenvironmental action: Attitude formation toward emergent attitude objects. *Journal of Applied Social Psychology, 25*, 1611–1636.

Thompson, S., & Barton, M. (1994). Ecocentric and anthropocentric attitudes toward the environment. *Journal of Environmental Psychology, 14*, 149–157.

Unger, R., & Crawford, M. (1996). *Women and gender: A feminist psychology*. New York: McGraw-Hill.

Van Liere, K., & Dunlap, R. (1978). Moral norms and environmental behavior: An application of Schwartz's norm-activation model to yard burning. *Journal of Applied Social Psychology, 8*, 174–188.

Van Liere, K. & Dunlap, R. (1980). A review of studies that measured environmental attitudes and behaviors. *Environment and Behavior, 11*, 22–38.

Widegren, O. (1988). The new environmental paradigm and personal norms. *Environment and Behavior, 30*, 75–100.

Wilkinson, S., & Kitzinger, C. (Eds.). (1996). *Representing the other: A feminist and psychology reader*. Thousand Oaks, CA: Sage.

Williams, J. E., & Best, D. L. (1990). *Sex and psyche: Gender and self viewed cross-culturally*. Newbury Park, CA: Sage.

Wolkomir, M., Futreal, M., Woodrum, E., & Hoban, T. (1997). Substantive religious belief and environmentalism. *Social Science Quarterly, 78*, 96–108.

Zelezny, L. C., & Yelverton, J. A. (2000, April). *Feminine identity, collectivism, and environmental attitudes and behaviors*. Paper presented at the meeting of the Western Psychological Association, Portland, OR.

LYNNETTE ZELEZNY is an Assistant Professor of Psychology at California State University, Fresno, where she teaches courses in environmental psychology, applied social psychology, statistics, and research methodology. Her research interests are related to environmentalism, gender, and minority issues. Recent publications have focused on the effectiveness of environmental education, translational action research in minority mental health, and cross-cultural research on environmental attitudes and behaviors. She is the author of *Methods in Action* (1999, Wadsworth) and numerous publications on environmentalism.

POH PHENG-CHUA is a doctoral student in social psychology at The Pennsylvania State University. Her research interests are in the psychology of stigma, ethnic identity, and minority issues.

CHRISTINA ALDRICH is a doctoral student in social psychology at The Claremont Graduate University. Her research interests are in psychology of women.

Journal of Social Issues, Vol. 56, No. 3, 2000, pp. 459–474

Models of Justice in the Environmental Debate

Susan Clayton*

The College of Wooster

Justice has become important in public and private consideration of the environment, but a number of different ways of operationalizing justice can be seen. Previous literature suggests that principles stressing responsibility and the public good are more common than need and equity in thinking about environmental issues. The results from two questionnaire studies, presented here, confirm that environmental justice—responsibility to other species and to future generations, and the rights of the environment—emerges as the most highly rated consideration in resolving environmental conflicts and that this factor is distinct from traditional procedural and distributive justice factors. Highlighting the individual or the collective makes different justice principles salient but that the effect depends on one's original position.

Justice has become a significant part of the public discourse over environmental issues. Many antienvironmental organizations (classified as such by Greenpeace; Deal, 1993) make explicit reference to justice in their titles: "Institute for Justice," "Fairness to Landowners Committee"; on the other side, the Sierra Club Legal Defense Fund recently changed its name to "Earthjustice Legal Defense Fund." Why has justice become an environmental issue? Three reasons can be identified. First, the relevance of justice depends on the ways in which one thinks about the resource. Justice becomes more salient, for example, under conditions in which a desired resource is scarce (Lerner, 1981). Our increasing consciousness that some resources are not renewable, within a meaningful time frame, has made us more aware of the ways in which those resources are distributed. It is also the case that the environment is a domain invested with moral significance and distinctive values for many, again raising the importance of justice

---

*Correspondence concerning this article should be addressed to Susan Clayton, The College of Wooster, Wooster, OH 44691 [e-mail: sclayton@acs.wooster.edu].

concerns (see Kempton, Boster, & Hartley, 1995; Stern & Dietz, 1994; Van Liere & Dunlap, 1978).

Second, distributive justice is highlighted when environmental benefits or hazards are linked to group identities. Along these lines, the racial disparities in exposure to environmental toxins described by the Commission for Racial Justice in 1987 are primarily responsible for introducing the term "environmental justice" to public discourse (see Bullard, 1990; Bullard & Johnson, this issue). United Nations meetings on environmental issues may have contributed to an awareness of national inequities in consumption of environmental resources and in expectations for environmental sacrifices.

Finally, an increased awareness that humans can and do have a serious and lasting impact on the natural environment has led to an increased perception of responsibility and of moral obligations, at least among some segments of the population (Schwartz, 1975).

It is significant that environmental issues are being evaluated in terms of justice. Researchers have begun to find that perceived justice is a good predictor of environmental attitudes, often better than self-interest (e.g., Kals, 1996), and that it is an important factor in the successful resolution of an environmental conflict (e.g., Lofstedt, 1996). There are multiple definitions of justice, however, with room for a number of factors to bias the definition in a self-serving way. Although we tend to assume, when invested in a situation, that one resolution is clearly the best and fairest, other constituents may see an alternative outcome just as clearly as being superior. So there is room for disagreement, and in the minds of the relevant participants this disagreement may be phrased not as "my interests" versus "their interests" but as "the side of right and good" versus "the side of expediency, greed, and immorality." This has obvious consequences for people's willingness to compromise.

The Forms of Environmental Justice

The Justice of the Marketplace

Probably the most familiar way of thinking about distributive justice, at least in the United States, is to frame it in terms of the marketplace, suggesting that the fair way to deal with natural resources is to sell them. This definition of environmental justice is already represented in public policy. The Chicago Board of Trade now includes a market in pollution credits, which can be bought (allowing industrial plants to pollute above a specified level) or sold (by plants that pollute less than their allowed amount; Power & Rauber, 1993). Individuals also participate in marketplace justice, when cities charge them for the number of trash bags they put out to be collected or when they pay significant user fees for access to national parks.

Equality

More recently, environmental justice has been framed in terms of equality. From this perspective, the current state of affairs is unjust because some people and countries consume far more of our environmental resources than others, and some people and countries are affected by environmental pollution to a far greater extent than others. The standard use of the term "environmental justice" in current discourse arose in response to the fact that poor, rural, and Black communities in the United States are disproportionately chosen as sites for the disposal of toxic waste because their lack of political power has made them less likely to mount effective resistance (Bullard, 1990; U.S. General Accounting Office, 1983) and implies the struggle to ensure that no groups, particularly minority groups, suffer disproportionately from the effects of environmental degradation.

Procedural Issues

Procedural justice places the focus on the fairness of the process by which goods are allocated and decisions made and particularly emphasizes the opportunity for all interested parties to participate in the decision process (Lind & Tyler, 1988). A common complaint from all parties is that environmental regulations are developed unfairly, without giving opposing sides the opportunity to participate. The environmental movement has specific procedural criticisms. One is that the rules that are on the books are not being enforced. Another is that big companies exert disproportionate influence on policymaking. A third procedural criticism relates to the idea of "environmental justice" and charges that various involved parties, usually those without much social power, have been excluded from participation in decisions about environmental matters that will affect them. Anti-environmental arguments also include procedural criticisms, generally that environmental policies are formulated by elitists and exclude the voices of the ordinary citizen.

Rights

Many environmental conflicts are framed in terms of rights. Rural landowners, along with timber and mining companies, have been filing lawsuits based on the idea that their "property rights" have been violated by environmental restrictions on the ways in which private property can be used. Environmental organizations have also focused on rights in an attempt to expand the concept, for example, by establishing environmental protection as a basic human right (Parker, 1991) or by allocating rights to entities that normally might not be considered: future generations, for example, or nonhuman entities (animals, species, ecosystems).

Responsibility

A last way of thinking about environmental justice invokes responsibility, which Schwartz (1975) has described as implicated in "the justice of need." Individuals are admonished to recognize the obligation to care for someone or something else. This argument is often seen in the environmental literature and particularly in the writings of Wendell Berry (1981). Berry takes seriously the idea that humans are meant to be stewards over the earth and argues that this means taking care of natural resources rather than taking them for granted. Similarly, the preface to the Sierra Club handbook for environment activists states that "each person must become ecologically responsible" (Mitchell & Stallings, 1970, p. 12). The appeal to responsibility can also be seen on the other side when environmental policies are criticized for an inattention to, or lack of concern over, the cost to people and communities whose jobs and ways of life are threatened.

Evaluating the Models

My research and that of others has addressed the question of what justice principles are preferred in resolving environmental conflicts. With some exceptions, unsurprising given the differences in domain as well as subject pool, we have found similar results. The highest ratings are given to specifically environmental concerns, like the rights of nature and concerns for future generations. Procedural justice and societal concerns like managing natural resources for the public good and the rights of the public to environmental resources also receive high ratings. Distributive principles like need and equity are rated lowest, with an economic model in which people pay for what they use consistently receiving among the lowest ratings of all (Clayton, 1996; McCreddin, Syme, Nancarrow, & George, 1997; Syme & Fenton, 1993; Syme & Nancarrow, 1992).

A slightly different question is whether particular models of justice are more congenial to different sides of the argument over environmental policy. Appeals to responsibility seem more successful in an environmental domain than in the service of other arguments (Clayton, 1994), and principles of responsibility have been articulated by citizens discussing international pollution (Linneroth-Bayer & Fitzgerald, 1996), but this is likely to vary according to the direct responsibility of the person answering the question. Opotow and Clayton (1998) recently surveyed material from antienvironmental organizations and found three principles to predominate: First, property rights are paramount. Second, regulations are unfairly imposed by government, encouraged by environmental extremists, that work to the detriment of the powerless average American. This makes reference to the concept of procedural justice. Third, market mechanisms are the fairest way to protect the environment. The types of justice referred to by these antienvironmental groups reflect a philosophy of individualism rather than interdependence, of rights rather than responsibility.

In contrast are the findings from an earlier analysis of fund-raising letters from environmental organizations (Clayton, 1991). Here, a principal emphasis was on the need to enforce existing standards or laws and thus ensure procedural justice. Another, less common theme was the need to expand our moral community (to include nature or future generations). Some unusual examples of "rights" were cited, including the "right of future generations to marvel at the Arctic National Wildlife Refuge" and the "right to a clean, safe, and healthy environment." A number of organizations made implicit or explicit appeals to responsibility.

Cvetkovich and Earle (1994), in a field study of an environmental conflict, reported that supporters of a wildlife protection ordinance were more likely than opponents to make reference to themes of equality or communal sharing. Dunlap and Van Liere (1984) found that support for private property rights was strongly and negatively associated with environmental concern. In an earlier laboratory study (Clayton, 1994), I found that although arguments based on procedure received equal agreement in an environmental and an antienvironmental condition, arguments based on responsibility or equality of suffering received significantly higher levels of agreement in the environmental than in the antienvironmental case.

In sum, research is accumulating to suggest that marketplace justice is more likely to represent the antienvironmental position and that themes of equality and responsibility are more common in environmental arguments. Rights and procedural issues are popular with both sides, though they are likely to be conceptualized differently. One way of describing the distinction between different models of justice is whether they are concerned with justice for the individual (microjustice) or justice for the larger society (macrojustice). Environmental groups seem more likely to utilize arguments based on macrojustice principles, whereas antienvironmental organizations find microjustice principles more compatible. (I have discussed the philosophical reasons for this elsewhere; see Clayton, 1994, 1996, 1998.)

In an attempt to strengthen the evidence for the preference for particular models of justice in evaluating environmental issues, as well as to explore further the link between a group orientation and environmentalism, I conducted two studies. In the first, students completed a group identification scale; in the second, an experimental manipulation was used to induce a high or low level of group identification. In both studies, participants read descriptions of environmental conflicts and then rated the importance of a number of justice principles in resolving each conflict. Information about gender and environmental attitudes was also recorded.

Study 1

Method

Participants and procedure. Data were collected from 93 college students: 33 men, 56 women and 4 who did not specify their sex. Each student read two scenarios describing environmental conflicts from either a proenvironmental or an antienvironmental perspective. The order of the scenarios and the condition of each scenario were both varied independently. After each scenario the students were asked to rate the importance of 12 justice principles in resolving the conflict. Finally they were asked to complete a group identification scale and the Environmental Attitudes Scale (EAS). The EAS was developed by Thompson and Barton (1994) to distinguish among different attitudes toward the environment: ecocentric, or a valuing of nature for its own sake; anthropocentric, or a valuing of nature for the benefits it can provide people; and apathetic. Upon completing all the materials students were carefully debriefed. The debriefing emphasized the fact that environmental issues could be looked at from a number of perspectives and that participants should not consider the position expressed in the materials they read to be definitive. Students received course credit for their participation.

Materials. One of the scenarios concerned the ability of the government to restrict private land use, and the other concerned logging of old-growth forests. The proenvironmental and antienvironmental perspectives were written in as similar a way as possible. Each scenario made reference to the opposing position, so that students' responses would reflect one position's being more salient than another rather than ignorance of the other position.

Results

Although the pattern of findings was largely the same for both scenarios, there were a number of effects that were significant only for one of the two. Since my interest was in finding ways of thinking about environmental issues that were general, effects will only be discussed if they were significant for one scenario and at least approached significance for the other.

Ranking of principles. Table 1 displays the justice principles in rank order according to mean rating of importance for each scenario. Note that collectivist or societal ideals and environmental principles were ranked very high, with procedural principles only slightly lower. Traditional principles of distributive justice, such as merit or economic principles, were ranked fairly low. This is consistent with the research indicating that environmentalism is very popular and that the rule of the marketplace is considered inappropriate for environmental issues. The ranking of principles was very similar across the two scenarios, suggesting consistency in the way justice is viewed for different environmental conflicts.

Factor analysis. To see if the 12 principles rated could be reduced to a smaller number of factors, each set of 12 ratings (those following the property scenario and those following the logging scenario) was subjected to a principal-components analysis. In each case, four factors with an eigenvalue greater than 1 were identified. Correlations between the factor scores for the two scenarios confirmed that three of the factors were substantially similar (with correlations of .55–.69, all significant at $p < .001$) across scenarios. The fourth factor varied between scenarios and was not examined further.

Among the three retained factors, one loaded primarily on the following items: "Allowing people to continue to use resources they have been using," "Making sure people get what they need," and "Making sure people get what they have

Table 1. Ranking and Factor Loadings for Justice Principles, Study 1

Justice principle[a]	Factor loadings[b]		
	Distributive	Environmental	Procedural
Logging scenario			
Responsibility to future generations		.65	
The rights of the environment		.80	
The chance for everyone to have a say			.86
Allocating the resources so that they will be used efficiently			
Responsibility to other species		.87	
Managing natural resources for the public good		.43	
The equal treatment of all groups			.84
The rights of the public to enjoy the environment	.42	.50	
Making sure people get what they have worked for	.70		
Making sure people get what they need	.78		
Making sure people pay for what they use			
Allowing people to continue to use resources they have been using	.73		
Private land scenario			
Responsibility to future generations		.45	
Responsibility to other species		.89	
The chance for everyone to have a say			.65
The rights of the environment		.87	
Allocating the resources so that they will be used efficiently			
The equal treatment of all groups			.89
Managing natural resources for the public good			
The rights of the public to enjoy the environment			.52
Making sure people get what they have worked for	.81		
Making sure people get what they need	.69		
Making sure people pay for what they use	.73		
Allowing people to continue to use resources they have been using	.68		

[a]Principles are listed in order of rated importance, from most to least important.
[b]Only factor loadings of .40 or greater are listed.

worked for." I called this the distributive justice factor, since it specifies who should receive what resources.

A second factor loaded principally on two items: "The rights of the environment" and "Responsibility to other species," with a substantial loading also on "Responsibility to future generations." I called this the environmental justice factor.

The third factor loaded on "The equal treatment of all groups" and "The chance for everyone to have a say." I called this the procedural justice factor. Factor loadings are included in Table 1.

Main effects. Simply presenting the conflict from a proenvironmental or an antienvironmental perspective had no effect. Ecocentrism predicted weight given to the environmental justice factor ($rs = .42$ and $.52$ for the two scenarios; $ps < .001$) Environmental apathy predicted less weight to the environmental justice factor ($r = -.40, p < .001$, for both scenarios) and more weight to the distributive justice factor ($rs = .24, p < .05$, and $.30, p < .01$, respectively). Anthropocentrism predicted a slight increase in support for distributive justice (significant only for the logging scenario, $rs = .22$ and $.16$). There were no consistent effects of gender or level of group identification on the factor scores.

Interaction effects. To allow for an examination of the interaction between subject variables and the independent variable, the former were dichotomized according to a median split. The environmental principles showed an interaction between experimental condition and group identification: In the nonenvironmental condition, people high in group identification rated these principles as more important than did people low in group identification, but in the environmental condition low group-identified people rated them higher than did high group-identified people. Means are shown in Table 2. In general, experimental condition made more of a difference in the ratings of people low in group identification than in those of people high in group identification. It may be that environmentalism has become part of the collective ideology to such an extent that high group-identified people show less variability in their endorsement of these principles than do those low in group identification.

Table 2. Mean Ratings of Environmental Justice Factor by Condition and Level of Group Identification, Study 1

	Group Identification	
Logging scenario	Low	High
Proenvironment condition	.43 (17)	−.02 (26)
Antienvironment condition	−.27 (28)	−.06 (18)
Parks scenario	Low	High
Proenvironment condition	.32 (25)	−.27 (18)
Antienvironment condition	−.30 (19)	.03 (26)

Note. Cell *n*s are listed in parentheses after each mean.

There were no other significant interactions with either experimental condition or group identification.

There was a similar pattern of interactions between participant sex and each of the environmental attitude dimensions on the environmental justice factor: The interactions between sex and environmental apathy, sex and anthropocentrism, and sex and ecocentrism were all significant. Means are shown in Table 3. People higher in apathy or anthropocentrism or lower in ecocentrism tended to have lower ratings of environmental justice, but the difference due to this attitudinal classification was greater for men than for women. Environmental attitudes generally are stronger for women than for men (see Zelezny, this issue), and this was true in the present study: Women were significantly lower than men in environmental apathy and higher in ecocentrism at a level approaching significance ($p < .10$). Thus women's tendency to refer to environmental justice principles may be less dependent on other factors than is the case for men.

Discussion

This study seemed to confirm the existence of a distinct "environmental justice" defined not so much by a particular philosophical perspective (e.g., equality or rights, individual or group level) as by the inclusion of remote entities, such as the environment or future generations, in one's consideration of a just

Table 3. Mean Ratings of Environmental Justice Factor by Participant Sex and Environmental Attitudes, Study 1

	Male	Female
Logging scenario		
Ecocentrism		
Low	−.69 (18)	−.06 (24)
High	.33 (15)	.22 (32)
Anthropocentrism		
Low	.31 (14)	−.04 (31)
High	−.62 (19)	.17 (25)
Environmental apathy		
Low	.60 (11)	.26 (37)
High	−.65 (22)	−.21 (19)
Parks scenario		
Ecocentrism		
Low	−.75 (18)	−.31 (24)
High	.72 (15)	.26 (31)
Anthropocentrism		
Low	.54 (14)	.06 (30)
High	−.54 (19)	−.05 (25)
Environmental apathy		
Low	.89 (11)	.25 (36)
High	−.56 (22)	−.44 (19)

Note. Cell *n*s are listed in parentheses after each mean.

resolution to a conflict. Results also supported the hypothesis that macrojustice principles are seen as more relevant than microjustice to an environmental issue.

The group identification scale did not demonstrate the expected pattern of results. It was not consistently related to preference for any particular type of justice principle, not even those principles that stressed the public good or public rights. It may be that the effect of group identification as tapped by this measure was not, as I had predicted, to lead to a greater valuing of group welfare, but rather to lead to a more consistent endorsement of those values that are part of the group ideology, namely, environmentalism. To examine the possible impact of group identification further, I designed a second study in which it would be manipulated as an independent variable.

Study 2

Method

Participants and procedure. Sixty-six undergraduate students (38 women, 28 men) participated in the study in return for extra credit in psychology or sociology courses. Each participant was arbitrarily assigned to either the low ($n = 34$) or the high ($n = 32$) group identity condition. In the low–group id condition, participants were told,

> We are interested in exploring the attitudes and opinions of different individuals toward environmental issues. To help us in understanding your responses, please take a few minutes to think of all the ways in which you are unique or distinctive among other College of Wooster students. This may include different background characteristics, beliefs, values, or interests. List these distinctive features below.

Participants in the high–group id condition were told,

> We are interested in exploring the attitudes and opinions of College of Wooster students toward environmental issues. To help us in understanding your responses, please take a few minutes to think of all the ways in which you are similar to other College of Wooster students. This may include similar background characteristics, beliefs, values, and interests. List these commonalities below.

This manipulation was adapted from Trafimow, Triandis, and Goto (1991). Following this task, participants read two scenarios describing environmental conflicts. The order in which these scenarios appeared was counterbalanced. After each scenario, participants were asked to rate the importance of 13 justice principles in solving the conflict. (The list of principles is similar, but not identical, to that in Study 1.) Finally, participants completed the EAS.

Materials. Participants each read two scenarios. One concerned the ability of the government to place restrictions on the way private landowners may develop their land. The second involved whether national parks should be made accessible to the public or left in their natural state. In each scenario, both pro- and anti-environmental positions were presented in a neutral manner.

Results

Rating of principles. The ranking of the 13 justice principles is shown in Table 4. As can be seen, the most important principles again were the ones protecting the interests of what can be called "environmental" constituents: future generations, other species, and the environment in general. The next most important principle was procedure, the chance for everyone to have a say. Fairly close were some macrojustice principles, the equal treatment of all people and managing resources for the public good, but individual rights were also rated at about this level. Responsibility to other people came in a little lower, as did public rights to the use of the environment. The least important principles were distributive ones: allocating resources on the basis of need or work, or equally, with the principle that people should pay for what they receive obtaining the lowest ratings in both scenarios.

Table 4. Ranking and Factor Loadings for Justice Principles, Study 2

Justice principle[a]	Factor loading[b]		
	Distributive	Environmental	Procedural
Private land scenario			
Responsibility to future generations		.71	
The rights of the environment		.52	
Responsibility to other species		.82	
The chance for everyone to have a say			.83
Individual rights			
Managing natural resources for the public good			
The equal treatment of all people			.44
Responsibility to other people			
Making sure people get what they have worked for	.65		
The rights of the public to the use and enjoyment of environmental resources			
Making sure people get what they need	.62		
Making sure costs or sacrifices are equally distributed	.46		
Making sure people pay for what they use	.75		
Parks scenario			
The rights of the environment		.80	
Responsibility to future generations		.76	
Responsibility to other species		.75	
The chance for everyone to have a say			.80
Managing natural resources for the public good			
The equal treatment of all people			
The rights of the public to the use and enjoyment of environmental resources			
Individual rights			
Responsibility to other people			
Making sure people get what they have worked for	.67		
Making sure costs or sacrifices are equally distributed			
Making sure people get what they need	.40		
Making sure people pay for what they use	.79		

[a]Principles are listed in order of rated importance, from most to least important.
[b]Only factor loadings of .40 or greater are listed.

Factor analysis. The ratings of the 13 principles for both scenarios were entered into a principal-components analysis. Eight factors were obtained with eigenvalues greater than 1.0, but both a scree test and interpretability suggested retaining four factors (which together explained 56% of the variance). Three of the factors were similar across scenarios; a fourth, which could be called humanistic or person-centered, was fairly specific to the second scenario, and so is not of much interest in terms of discovering general principles of justice (it may relate specifically to attention to the needs of the disabled). Factor loadings for the three general factors are shown in Table 4. The first factor has the highest loading on the environmental principles. The second loads on the distributive principles. The third is almost entirely focused on the procedural justice principle, the chance for everyone to have a say.

Main effects. Again, women showed more environmental attitudes: There was a significant gender difference in ecocentrism, with women ($M = 52$) scoring higher than men ($M = 48.7$), and a tendency ($p < .10$) toward a parallel difference in environmental apathy, with men ($M = 19.3$) scoring higher than women ($M = 16.6$). There were no gender differences in any of the factor scores.

Ecocentrism was significantly correlated with the environmental factor ($r = .42, p < .01$) and with none of the other factors. Apathy was negatively correlated with the same factor ($r = -.45, p < .01$) and with no other factor.

There were no main effects of condition on any of the retained factor scores.

Interaction effects. Ecocentrism, anthropocentrism, and apathy were subjected to a median split and the resulting categories were used in analyses of variance to see whether they interacted with experimental condition. There were significant interactions between ecocentrism and experimental condition for the procedural justice factor and a trend toward an interaction for the distributive justice factor ($p = .10$). Means are shown in Table 5. People high in ecocentrism gave more importance to the procedural justice variable in the high–group id condition than in the low–group id condition, whereas the reverse was true for people low in ecocentrism; the opposite pattern was true for the distributive justice variable.

There was a similar interaction between experimental condition and subject gender on the distributive justice factor and a trend toward such an effect for the procedural justice factor ($p = .11$). Table 5 shows the means. Men rated distributive justice as being more important in the high–group id condition than the low–group id condition, whereas women showed the opposite pattern. The reverse was true for the procedural justice factor.

Table 5. Means for Interaction Effects on Distributive and Procedural Justice Factors, Study 2

Factor	Ecocentrism		Sex	
	Low	High	Men	Women
Distributive justice				
Experimental condition				
Low group id	−.17 (15)	.11 (17)	−.35 (11)	.30 (21)
High group id	.32 (17)	−.22 (17)	.47 (17)	−.20 (17)
Procedural justice				
Experimental condition				
Low group id	.19 (15)	−.40 (17)	.08 (11)	−.29 (21)
High group id	−.11 (17)	.30 (17)	−.19 (17)	.26 (17)

Note. Cell *n*s are listed in parentheses after each mean.

Discussion

There is further evidence here for an environmental justice factor. The factor analysis showed that this factor is distinct from the procedural and distributive concerns; the rankings of the different principles show that this factor is considered to be important in resolving environmental conflicts.

The experimental manipulation of collective orientation was influential in two ways. The first was an interaction between experimental condition and participant sex on the distributive justice factor and marginally on the procedural justice factor. The second was an interaction between experimental condition and ecocentrism on the procedural justice factor and marginally on the distributive factor. Since the pattern was the same for women and people high in ecocentrism as compared to men and people low in ecocentrism, and since women were higher in ecocentrism than men, the two variables may be tapping the same underlying dimension. I hypothesize that this dimension may reflect some level of group identification or moral community.

Distributive justice principles were rated as less important in the high–group id condition, as would be expected if the group id manipulation made macrojustice more salient, but only by women. It may be that men gave higher ratings to these principles in the high–group id condition because, for them, collective consciousness means thinking about other members of a group as individuals and so a proportional distribution is the best outcome. When they are thinking about themselves as distinctive they may be less thoughtful about outcomes to others overall. This interpretation is supported by the fact that men gave lower ratings than women for 19 of the 26 principles (usually *ns*), nonsignificantly higher in only 3, and about equal in the remaining 4.

For the procedural justice factor, the reverse pattern was obtained. People high in ecocentrism place more emphasis on procedural issues in the high–group id condition. This may be a function of the (environmental) type of scenario. When people high in ecocentrism are encouraged to think about the group while considering an environmental conflict, they may be more concerned with inclusion of other people, and hence procedure, but when they are thinking of themselves as distinctive they are more focused on nonhuman interests. For those low in ecocentrism, when they are thinking of themselves as distinctive they may be afraid that their own voices and the voices of people like them will not be included in environmental conflict resolution. For this reason they give procedural justice concerns more importance than when they are thinking about similarities with the group.

General Discussion

How can we achieve justice in environmental conflicts? Clearly, there are multiple justices to consider. This may mean that it is impossible to satisfy all justice concerns simultaneously. However, environmental policies should at least show awareness of the competing fairness issues. In the absence of open discussion, each justice argument may be compelling to those who have not thought about the complexity of the issue, and those whose concerns are not addressed will feel as if they have been excluded from the community of those relevant to the decision outcome. Public consideration of different arguments may increase the understanding of opposing positions and the legitimacy and ultimate acceptability of the proposed solution. An awareness of new factors, like the decreased availability of resources, or a shift to a long-term rather than a short-term perspective, may lead people to change their original position and arrive at a new acceptance of other people's perspective.

Both these studies support the idea of a distinct "environmental justice" emerging as an issue in resolving environmental conflicts. This may be sufficiently novel that not everyone will spontaneously consider its relevance but sufficiently popular that most will concede its legitimacy. Alternatively, some may reject the idea that nonhuman entities deserve moral standing or even that justice is relevant to environmental issues.

Perhaps the most important or the most overarching question with regard to environmental justice is the issue of justice for whom: How is the moral community defined? Do we include other species, future generations, fuzzy concepts like ecosystems? Do individual concerns automatically get overridden by the needs of the larger society or is there a way to preserve justice at both levels? The limits of the relevant community will affect the determination of all other issues, including distribution, procedure, rights, and responsibilities.

Ultimately, much of the discourse on environmental issues is about identity. The impact of participant sex in the above studies indicates that group identity matters. The perceived fairness of an environmental policy will depend on who is defined as a relevant member of the justice community: who is entitled to have his or her interests considered (cf. Opotow, 1990, 1994; Opotow & Weiss, this issue; see also Bullard & Johnson; Dunlap, Van Liere, Mertig, & Jones; and Schultz, all in this issue). Encouraging people to think at a collective level, though useful in bringing relevant aspects of environmental issues into perspective, will have a different impact depending on the groups of which they feel themselves to be members. Some people feel connected to the natural environment and may merely need to be reminded of the need to take responsibility for it. Others may be alienated from the environment and most successfully won over by arguments about the human benefits of environmental policies. A third group seems to exist who feel actively hostile to the efforts of environmentalists; for them, collective thinking may serve to highlight a sense of intergroup tensions between environmentalists and themselves.

Exploring the level of inclusion could be useful for both researchers and decision makers. One outcome of research into environmental justice could be further refinements in our definitions of what counts as full inclusion and whose interests need to be included, at least by proxy. The more voices included in the decision process, the more researchers can see the competing definitions of environmental justice that emerge. We should neither hope nor expect that social scientists can resolve the different concepts of justice into a single, "true" definition, but through awareness of the different definitions, both a justice theory and a reality that are more encompassing of different perspectives may be achieved.

References

Berry, W. (1981). *The gift of good land*. San Francisco: North Point Press.
Bullard, R. (1990). *Dumping in Dixie: Race, class, and environmental quality*. Boulder, CO: Westview Press.
Clayton, S. (1991, August). Environmental justice. Paper presented at the annual meeting of the American Psychological Association, San Francisco, CA.
Clayton, S. (1994). Appeals to justice in the environmental debate. *Journal of Social Issues, 50*(3), 13–27.
Clayton, S. (1996). What is valued in resolving environmental dilemmas: Individual and contextual influences. *Social Justice Research, 9*, 171–184.
Clayton, S. (1998). Preference for macrojustice vs. microjustice in environmental decisions. *Environment and Behavior, 30*, 162–183.
Cvetkovich, G., & Earle, T. (1994). The construction of justice: A case study of public participation in land management. *Journal of Social Issues, 50*(3), 161–178.
Deal, C. (1993). *The Greenpeace guide to anti-environmental organizations*. Berkeley, CA: Odonian Press.
Dunlap, R. E., & Van Liere, K. (1984). Commitment to the dominant social paradigm and concern for environmental quality. *Social Science Quarterly, 65*, 1013–1028.
Kals, E. (1996, July). Pollution control, justice, and responsibility: Who is willing to engage in the protection of air quality? Paper presented at the Fourth International Conference on Social Justice, Trier, Germany.
Kempton, W., Boster, J. S., & Hartley, J. A. (1995). *Environmental values in American culture*. Cambridge, MA: MIT Press.

Lerner, S. (1981). Adapting to scarcity and change: Stating the problem. In M. J. Lerner & S. C. Lerner (Eds.), *The justice motive in social behavior* (pp. 3–10). New York: Plenum.

Lind, E. A., & Tyler, T. R. (1988). *The psychology of procedural justice*. New York: Plenum.

Linneroth-Bayer, J., & Fitzgerald, K. B. (1996). Conflicting views on fair siting processes: Evidence from Austria and the U.S. *Risk: Health, Safety and Environment, 7*, 119–134.

Lofstedt, R. E. (1996). Fairness across borders: The Barseback nuclear power plant. *Risk: Health, Safety, and Environment, 7*, 135–144.

McCreddin, J. A., Syme, G. J., Nancarrow, B. E., & George, D. R. (1997, February). *Developing fair and equitable land and water allocation in near urban locations: Principles, processes, and decision making* (CSIRO Division of Water Resources Consultancy Report No. 96-60). Perth, Australia: CSIRO.

Mitchell, J. G., & Stallings, C. L. (Eds.). (1970). *Ecotactics: The Sierra Club handbook for environment activists*. New York: Simon & Schuster.

Opotow, S. (1990). Moral exclusion and injustice: An introduction. *Journal of Social Issues, 46*(1), 1–20.

Opotow, S. (1994). Predicting protection: Scope of justice and the natural world. *Journal of Social Issues, 50*(3), 49–63.

Opotow, S., & Clayton, S. (1998, August). *What is justice for citizens' environmental groups?* Paper presented at the International Conference for Applied Psychology, San Francisco.

Parker, V. (1991, Spring). Legal defense fund mounts major international effort to link human and environmental rights. *In Brief* [newsletter of the Sierra Club Legal Defense Fund], pp. 1, 7.

Power, T. M., & Rauber, P. (1993). The price of everything. *Sierra, 78*(6), 87–96.

Schwartz, S. (1975). The justice of need and the activation of humanitarian norms. *Journal of Social Issues, 31*(3), 111–136.

Stern, P., & Dietz, T. (1994). The value basis of environmental concern. *Journal of Social Issues, 50*(3), 65–84.

Syme, G., & Fenton, D. (1993). Perceptions of equity and procedural preferences for water allocation decisions. *Society and Natural Resources, 6*, 347–360.

Syme, G., & Nancarrow, B. (1992). *Perceptions of fairness and social justice in the allocation of water resources in Australia* (CSIRO Consultancy Report No. 92-38). Perth, Australia: CSIRO.

Thompson, S. G., & Barton, M. A. (1994). Ecocentric and anthropocentric attitudes toward the environment. *Journal of Environmental Psychology, 14*, 149–157.

Trafimow, D., Triandis, H., & Goto, S. (1991). Some tests of the distinction between the private self and the collective self. *Journal of Personality and Social Psychology, 60*, 649–655.

U.S. General Accounting Office. (1983). *Siting of hazardous waste landfills and their correlation with racial and economic status of surrounding communities*. Washington, DC: Government Printing Office.

Van Liere, K. D., & Dunlap, R. E. (1978). Moral norms and environmental behavior: An application of Schwartz's norm-activation model to yard burning. *Journal of Applied Social Psychology, 8*, 174–188.

SUSAN D. CLAYTON received her PhD from Yale University and is currently an Associate Professor of Psychology at the College of Wooster. She has published on topics related to justice and the environment in several books and in journals such as *Social Justice Research* and *Environment and Behavior*. With Susan Opotow, she coedited a previous issue of the *Journal of Social Issues* on "Green Justice: Conceptions of Fairness and the Natural World." An earlier book (with Faye Crosby), *Justice, Gender, and Affirmative Action* (1992) received an Outstanding Book Award from the Gustavus Myers Center for the Study of Human Rights.

Journal of Social Issues, Vol. 56, No. 3, 2000, pp. 475–490

Denial and the Process of Moral Exclusion in Environmental Conflict

Susan Opotow*

University of Massachusetts Boston

Leah Weiss

Annapolis, MD

Environmental issues present an urgent challenge throughout the world. Air, water, and land pollution continue at alarming rates and increasingly strain the Earth's capacity to sustain healthy ecosystems and human life. Although techno- logical and behavioral aspects of environmental conflict are often salient, this article contributes to the literature on environmentalism by examining moral orientations that underlie and fuel environmental conflict. The centerpiece of this article describes three kinds of denial in environmental conflict: (1) outcome severity; (2) stakeholder inclusion; and (3) self-involvement. Like intermeshed gears, these forms of denial actively advance the process of moral exclusion. The article concludes with implications of this analysis for theory and practice.

Environmental issues present an urgent challenge throughout the world. Air, water, and land pollution continue at alarming rates and increasingly strain the Earth's capacity to sustain healthy ecosystems and human life. Although marked improvements in environmental quality have been documented in the United States for certain pollutants, additional questions, concerns, and conflicts continue to arise over the current state of the environment locally, regionally, nationally, and internationally. For example, pedestrians and bus drivers dispute the prolonged idling of diesel engines in downtown business districts; a neighborhood challenges a proposed siting for an incinerator or power plant; the U.S. Environmental

*Correspondence concerning this article should be addressed to Susan Opotow, Graduate Program in Dispute Resolution, University of Massachusetts, 100 Morrissey Blvd., Boston, MA 02125-3300 [e-mail: susan.opotow@umb.edu].

Protection Agency battles industry groups over appropriate quality standards; and around the world, nations that were parties to the 1997 United Nations Framework Convention on Climate Change in Kyoto, Japan, struggle with marketplace forces to lower emissions of greenhouse gases.

Although technological and behavioral aspects of environmental conflict are often salient, this article describes subtle moral orientations that underlie the process of environmental conflict (cf. Clayton & Opotow, 1994). It offers an analysis of three forms of denial, discussing each in the context of air quality conflicts. These forms of denial advance self-serving moral justifications, exclusionary attitudes, structural change in the conflict, conflict escalation, and destructive conflict outcomes. Understanding environmental conflict, moral exclusion, and the contribution of denial to both is crucial for constructively managing conflict process.

Environmental Conflict

Conflicts result from behavioral or attitudinal incompatibilities as parties bring distinctive worldviews to conflicts that include their interests, positions, culture, beliefs, tactics, skills, needs, values, and perceptions of fairness (Deutsch, 1973). Conflicts can be obvious or hidden, constructive or destructive, and occur at smaller and larger contexts, from within individuals to international conflicts. Conflicts are often precipitated by as well as precipitate changes that include modifications in positions and attitudes, enlarging or narrowing of issues, and the mix of engaged stakeholders (Mather & Yngvesson, 1980–81). Morton Deutsch (1973) has identified the influential relationship between conflict processes and conflict outcomes. He proposes that positive conflict processes, such as cooperation, promote constructive conflict outcomes, whereas negative conflict processes, such as disrespect, distrust, and miscommunication, promote destructive conflict outcomes. Although all conflicts share these characteristics, *environmental conflicts* as a class have some distinctive characteristics. Promoting *environmentalism*—an environmentally protective stance and behavior—depends on an understanding of environmental conflict.

Large Scale

Environmental conflicts are unusually large scale and complex. They involve large numbers of people (often millions). These human stakeholders can differ substantially from each other in perceptions of risk, time horizons, and value and as well as in their access to power and political and economic resources (Susskind, 1981). Environmental conflicts involve complex systems that include regulatory bodies, proximate and distal parties, individuals and groups, and future stakeholders. In addition they involve nonhuman natural systems that remain incompletely understood (Susskind & Field, 1996). Because of the large numbers and diversity of human and nonhuman animate and inanimate stakeholders (e.g., rivers, etc.), environmental conflicts are often representation disputes debating who should be identified as valid spokespersons for specific positions and interests.

The Commons

Environmental conflicts concern shared resources (e.g., the watershed, the air, land use) and harms (e.g., pollution), evoking the dynamic of the commons. Garrett Hardin (1968) describes how shared space lends itself to environmental tragedy when the costs of overutilization accrue at the macro level but not the micro level. Adding sheep to the common benefits an individual sheep owner, but when widely adopted as a practice, leads to overgrazing and degradation of the common.

Specialized Knowledge

Environmental conflicts depend on the interpretation of scientific and technical data. This often results in two classes of stakeholders: those with and those without sufficient knowledge to interpret these data. Relatedly, environmental conflicts result from and shape public policies involving government agencies and formal regulatory decision-making processes. Some stakeholders are more knowledgeable about the complex array of regulatory mechanisms than others.

Throughout the article we will illustrate our analysis with examples of air quality conflict. This approach not only offers a coherent focus but also illustrates the three main characteristics of environmental conflict:

1. *Large scale*: The airshed is shared and needed by everyone. It also has limits on its use if it is to be maintained at a particular level or quality. Geographically the airshed is also large scale. Air pollutants can travel hundreds of miles; consequently, geographic areas that may not have a localized air pollution problem contribute to problems experienced downwind.

2. *Commons*: Air pollution results from micro and macro behaviors. Although major industrial sources contribute to the air pollution problem, so do an individual's activities. Adding a few small sources of polluting emissions yields nonenvironmental benefits (i.e., convenience) and minimal environmental impacts for an individual, factory, or community at the micro level, but as emissions accumulate from many sources, impact on the airshed can be substantial and harmful.

3. *Specialized knowledge*: Monitoring air quality and interpreting these data takes specialized equipment, knowledge, and skills. Although clean air is

critically important for public health, we neither see the air nor certain forms of pollution. Air pollution remains invisible to the public until it has reached unacceptable proportions, as in Los Angeles on a smoggy day. Long-term human costs of air pollution are also difficult for ordinary citizens to detect. They are, however, evident in chronic, debilitating respiratory and pulmonary disease. On a hospital pulmonary ward, where breathing can no longer be taken for granted, the effects of day-to-day air pollution are obvious, dramatic, and frightening.

In sum, the complex, large-scale dynamics of environmental conflicts make them difficult to analyze and resolve. They are particularly complex because they simultaneously involve multiple kinds of conflicts, including conflicts of interests and conflicts of values (Thompson & Gonzalez, 1997). The next sections describe the processes by which environmental conflicts are justified, progress, and escalate and can lead to destructive outcomes.

Moral Exclusion

The *scope of justice* is our psychological boundary for fairness. Norms, moral rules, and concerns about rights and fairness govern our conduct toward those inside our scope of justice (also called the *moral community*; Deutsch, 1985; Opotow, 1990). Our scope of justice is attuned to: Who and what counts? Who and what simply does not matter? The scope of justice emerges from three attitudes toward others: (1) believing that considerations of fairness apply to them, (2) willingness to allocate a share of community resources to them, and (3) willingness to make sacrifices to foster their well-being (Opotow, 1987, 1993). As Table 1 illustrates, these attitudes are consistent with environmentalism because they place the well being of the larger ecosystem above anthropocentic or personal concerns (cf., Merchant, 1980; Stern & Dietz, 1994), emphasize the interdependence of people and nature, view humans as only one of many parts of nature, and advocate decision making that considers the larger natural system in which humans are embedded. This perspective is evident in the Gaia hypothesis and its depiction of Earth as a single, interconnected system (Lovelock, 1979). It is also evident in philosophies that characterize inanimate natural objects, such as soil, land, rivers, and mountains, as part of the biotic community and therefore of concern when considering public policy (Leopold, 1949; Stone, 1974).

Moral concerns are relevant for those inside the scope of justice. *Moral exclusion*, in contrast, rationalizes and justifies harm for those outside, viewing them as expendable, undeserving, exploitable, or irrelevant. An exclusionary, anti-environmentalist perspective, exemplified by the "wise use" movement, asserts the preeminence of humans and values human economic and recreational activity over the well-being of the nonhuman natural environment.

Table 1. Moral Exclusion in Environmental Conflict

Inclusion in the scope of justice	*Environmentalism*
Believing that considerations of fairness apply to the other	Considering the natural world as well as personal or human well-being
Willingness to allocate a share of community resources to the other	Viewing nonhuman aspects of the natural world as an end rather than only as a means and entitled to resources
Willingness to make sacrifices to foster the other's well-being	Willingness to assume costs of environmental protection at the individual, group, community, or national level
Exclusion from the scope of justice	*Antienvironmentalism*
Believing that considerations of fairness do not apply to the other	Considering personal or human concerns as more important than the natural world
Unwillingness to allocate community resources to others	Viewing nonhuman aspects of the natural world as a means rather than as an end and therefore not entitled to resources
Unwillingness to make sacrifices to foster others' well-being	Unwillingness to assume costs of environmental protection

In environmental conflict, moral exclusion can be flagrant, but it can also be subtle and difficult to detect, for example when some stakeholders are not invited to such decision-making meetings as the drafting of regulations. Because moral exclusion can be invisible when shared social convention supports it, it can be difficult to detect in our own thinking and within our own culture as well. However, moral exclusion has characteristic symptoms that offer evidence of its presence (Opotow, 1990). Recognizing these symptoms can alert us to the presence of moral exclusion and kindle awareness of destructive conflict processes it can foment. An analysis of the symptoms of moral exclusion in air, land, and water conflicts indicates that these symptoms can be grouped into three kinds of denial that can expedite exclusionary perceptions, beliefs, and behavior (Opotow, Weiss, Lemler, & Brown, 1997).

A Typology of Denial in Environmental Conflict

Denial is "a defense mechanism consisting of an unconscious, selective blindness that protects a person from facing intolerable deeds and situations" (Corsini, 1999, p. 263). Denial is a form of selective inattention toward threat-provoking aspects of a situation to protect a person from anxiety, guilt, or other ego threats. Denial is a common and normal way of coping with problems and conflicts. Although it can promote healthy functioning, denial can also block attention to potential dangers to well-being. Psychoanalytical theories of denial focus on the individual and interpersonal relations. Although other disciplines also consider denial and self-deception, their focus includes the role of denial in social convention and at larger, societal levels of analysis (Weiss, 1997).

Evolutionary biologists view self-deception as wired into human makeup because of its adaptive value. Krebs, Denton, and Higgins (1988) propose that in our divided consciousness, in which the two brain hemispheres mediate contradictory behaviors, the left may "misinterpret and distort the knowledge possessed by the right" (p. 109). Positive effects of denial are evident in the *placebo effect*, "false beliefs about the connection between behaviors and consequences" (p. 129), associated with positive mental states and better healing after surgery or disease. The placebo effect is a form of denial that is adaptive because it stimulates motivation for persevering in the face of adversity.

Consistent with psychoanalytic theory, moral philosophers note that people tend to commit to well-supported ideas and avoid those that are painful. Denial arises when one deliberately maneuvers to avoid (1) accepting a new belief, (2) the pain of resisting a well-established belief, and (3) the pain involved in the belief that one is self-deceiving. They hypothesize that on some level self-deceivers know that they are self-deceiving. Thus, denial is the dialectical interplay of knowledge and will (Mullen, 1995).

Social theory views denial on a large scale as "the emergence of social amnesia, which makes and remakes society" (Jacoby, 1975, p. 4). Social amnesia allows people, individually and collectively, to cope with pain and tragedy, making it more concrete and palatable, even assigning it social value through any number of routes (e.g., reification and commodification). Social amnesia allows us to transform others from what they are to what we want them to be. This allows us to deal with, consume, or exterminate them, particularly when they pose a threat to oneself (Weiss, 1997).

As we will describe, denial not only has important functions in personal and social conflict but also in environmental conflict. Although the literature on environmental conflict has not heretofore examined its role, an understanding of denial in such conflicts is clearly critical. The next sections describe three forms of denial as interconnected "working parts" of moral exclusion, fostering exclusionary perceptions about the situation, the other, and oneself. Consequently, they justify behavior that influences the course and outcome of environmental conflicts.

Denial of Outcome Severity

Denial of outcome severity conceals environmental harms that accrue to oneself, others, and the nonhuman natural environment. As Table 2 indicates, denial of outcome severity is evident in such symptoms of moral exclusion as minimizing injurious outcomes resulting from environmental practices and policies, invoking different levels of environmental harm as acceptable for different social categories, asserting that exposure to or injury from harms is an isolated event rather than ongoing, and disavowing deteriorations in physical conditions. Denial of outcome severity depends on selective distortions of harms and data.

Table 2. Symptoms of Moral Exclusion

Denial of outcome severity	
Double standards	Invoking different levels of harms (e.g., pollution) as acceptable for different social groups
Concealing effects of harmful outcomes	Disregarding, ignoring, distorting, or minimizing injurious outcomes resulting from environmental practices
Temporal containment of harm	Asserting that exposure to or injury from harms is an isolated, unlikely event rather than routine and/or chronic
Reducing moral standards	Asserting that one's harmful behavior is proper while denying concerns for others
Utilizing euphemisms	Masking and sanitizing harmful outcomes with palliative terms, especially to disavow a deterioration in environmental conditions
Denial of stakeholder inclusion	
Biased evaluation of groups	Making unflattering between-group comparisons that bolster one's own ideological position or sense of superiority at the expense of others; emphasizing negative attributes of others while emphasizing positive attributes of oneself to discredit, exclude, and trivialize "outsiders'" interests, knowledge, or stake in a conflict compared with one's own
Condescension and derogation	Regarding other stakeholders with disdain and denigrating them
Dehumanization	Denying other stakeholders' entitlements to resources as well as denying their humanity and dignity
Fear of contamination	Perceiving contact or alliances with other stakeholders as posing a threat to one's position, credibility, or well-being, while denying the benefits that within-group diversity can offer
Normalization and glorification of violence	Glorifying and normalizing violence; viewing violence as an effective, legitimate, or even sublime form of human behavior while denying the potential of violence to damage people, the environment, relationships, and constructive conflict resolution processes
Denial of self-involvement	
Victim blaming	Displacing blame on those harmed
Deindividuation	Believing one's contribution to an environmental problem is undetectable
Diffusing responsibility	Denying personal responsibility for an environmental harm by seeing it as the result of collective rather than individual decisions and actions
Displacing responsibility	Identifying others, such as higher authorities, as legitimate decision makers responsible for environmental harms
Self-righteous comparisons	Casting oneself as environmentally "clean" and blameless in comparison to "dirty," irresponsible, or reprehensible others

Disbenefits. Degree of harm and degradation, often called *disbenefits* in the air regulatory field, are a central issue debated in environmental conflicts. Disbenefits include harms accruing to people and nonhuman environmental entities over time, including damage to air, water, or land commons at every level of analysis: individual, community, region, and so on. Because stringent environmental standards mandated by law do not necessarily promote stringent standards of compliance or enforcement, disbenefits are often debated phenomenologically, as regulatory issues, or as differential outcomes to particular groups.

Conflicts concern disbenefits, for example, when they focus on extent or severity of smog. Denial of disbenefits was apparent during negotiations of the Ozone Transport Assessment Group (OTAG). OTAG, comprised of 37 states, convened for 2 years (1995–1997) to identify a control strategy to reduce transported ozone across the eastern United States from upwind to downwind states (Weiss, 1996). During that process, several upwind states denied the well-documented nature of regional ozone transport and posited that only areas with an already-identified ozone problem should be required to control their emissions. As one representative of an upwind state stated:

> I am not so sure that we need long-range ozone-transport control because we are not seeing the ozone and its precursors transported over long distances, such as the Mississippi River to the East Coast. . . . If we are going to control ozone, it should probably be on a smaller scale. ("OTAG result: Severe Nox emissions cut," 1996, p. 8)

This position implies that the dirtiest power plants, many located in rural Midwestern areas that do not yet have high pollution levels, should be excused from controls. These stakeholders denied not only outcome severity but also the benefits of controls for local populations in the immediate vicinity of these power plants.

Because the disbenefit argument can bolster any side of an environmental dispute, it is important to understand it in context. During the OTAG negotiations, some stakeholders who denied airshed pollution severity nevertheless used the disbenefits argument to justify their unwillingness to control their own emissions. They presented evidence indicating that certain emission controls that would provide significant regional benefit would yield local disbenefits. Thus, an emphasis on localized disbenefits can sidestep an analysis of the larger picture and the relative, net benefits on a regional level. In this case, further review of the evidence offered revealed that disbenefits were very localized and for the most part at levels so low that they would rarely trigger violations of the air standard.

Advocacy science. As the previous example illustrates, environmental data play a key role in assessing and asserting outcome severity. Science is a tool, and how that tool is used depends on who is wielding it. Although science is universally acknowledged as an appropriate basis for making environmental policy, science is not monolithic and few facts are indisputable. Because facts are filtered by values (Stern & Dietz, 1994), there is a tension between the objective and subjective in science.

Decisions about the focus and methods of scientific investigation, kinds of data collected, kinds of analyses employed, interpretations of the data, and predictive reliability of those data are subject to interpretation and debate. As a result, scientific methods and findings often fuel rather than resolve environmental controversy. Proponents of a particular course of action can selectively use scientific findings to support their own beliefs and goals while denying the importance of underlying values and interests shaping their interpretation of scientific findings. During OTAG negotiations, computer modeling of smog had been initially scorned by some industry groups. Over time they embraced these models once they learned how to use the model and its results to their advantage.

Science can also be used as a tactic to block change or gain strategic advantage. Stating "we need more time to study the problem" not only utilizes science as a stumbling block but implicitly denies that harm can accrue from inaction. Finally, data are central to environmental conflict when some stakeholders challenge the assumptions of predictive data as inadequate or erroneous. When the data are overtly disputed, trust, communication, and commonality of goals are implicitly debated as well.

Denial of Stakeholder Inclusion

Although environmental conflicts ostensibly concern physical resources, much energy and acrimony questions the legitimacy of particular stakeholders. The participation of some stakeholders is mandated by regulatory bodies, whereas other stakeholders self-identify as legitimate because of their concerns. The legitimacy, reasonableness, and urgency of their concerns, however, often spark between- and within-group conflict. As Table 2 indicates, denial of stakeholder inclusion is evident in such symptoms of moral exclusion as unflattering between-group comparisons that bolster one's own ideological position or sense of superiority at the expense of other stakeholders; biased evaluations ignoring positive attributes of other stakeholders while ignoring negative attributes of oneself; trivializing other stakeholders' interests, knowledge, or stake in an issue when compared with one's own; regarding other stakeholders with disdain and disparaging or denigrating them; denying the entitlements of other stakeholders to resources, as well as denying their humanity and dignity; perceiving contact or alliances with other stakeholders as posing a threat to one's position, credibility, or well-being, while denying benefits that within-group diversity can offer; and glorifying and normalizing violence. These symptoms are conspicuous when labeling others as "outsiders" or "extremists."

Outsiders. In environmental disputes, parties often denigrate and label each other. Labeling can cast the other as an "outsider" with an identity fitting the labeler's agenda. In air quality disputes, regulators are seen by automobile manufacturers as "ecofreaks" who want to stop the driving public by adopting standards-forcing technology to introduce zero-emitting vehicles into the market. The same regulators are labeled "pinheaded bureaucrats" by citizens from another perspective who fear pollution and are unsatisfied with reasons given for siting a power plant in their community. Automobile manufacturers and oil companies are perceived by environmentalists as "foot-dragging big businesses" trying to make a buck while resisting the development and production of low-emission vehicles and legislation mandating cleaner fuels.

Extremists. Categorizing other stakeholders as "extreme" because of their lifestyle, ideology, or conflict resolution tactics marginalizes their concerns and potential to contribute constructively to the environmental dispute. Environmental conflicts are typically multiparty disputes yielding allies that cross traditional political boundaries. Unlikely and unstable coalitions can form that include a range of stakeholders from conservative to radical positions. Moderate stakeholders can fear extremists on their own side, seeing them as hindering public support, distracting from more important issues, and making compromise inordinately difficult. Moderates therefore justify excluding stakeholders with more extreme positions as an expedient facilitating conflict resolution. Moderates on opposite sides of an environmental conflict may share more beliefs and expectancies about the dispute resolution process than they share with extremists on their own side.

Including or excluding particular stakeholders changes the coalition strength supporting or opposing particular positions. The ideological mix of stakeholders on a particular side of a conflict can indeed influence public and political support as well as determine the kinds of trade-offs that may acceptably resolve a conflict. Therefore, stakeholder exclusion and inclusion influences the nature, course, and outcome of environmental conflicts.

Although within-group exclusion of "extremists" may be pragmatic, it risks losing the most distilled and thorny aspects of an environmental conflict. So-called extremists may hold the key to durable resolution because they raise issues that won't go away. Denying the concerns of extremists risks forging an easier, faster, but less enduring agreement unleavened by diversity of perspectives. In the U.S. Environmental Protection Agency's environmental regulatory negotiations ("reg-negs"), it is often difficult for some stakeholders to procure a seat at the table. As a consequence, many of those negotiated settlements have disappointed environmentalists, industry, and state regulators. Thus, denial of the legitimacy or relevance of extremists' concerns may work, but only in the short run. In the long run, extremists' commitment, persistence, principle-based positions, willingness to sacrifice, and access to media attention can ultimately change public opinion.

Stakeholders labeled extremists and excluded from mainstream dispute resolution forums may utilize warfare-like tactics to achieve ends they deem important. Because of the attention that dramatic actions can garner, they can achieve public recognition, engage the public in environmental issues attitudinally and behaviorally, and ultimately, influence the conflict process. Environmental groups once considered extreme, such as Greenpeace and Earthfirst!, continue to gain mainstream acceptance by successfully challenging the acceptability of current practice. The Public Interest Research Group's (PIRG's) campaign to identify and shut down old power plants across the nation, at one time a radical idea, received significant political support from Democratic and Republican gubernatorial candidates in recent elections.

Although extreme positions are not inevitably connected with violence, those who self-identify as extremist or approve of extreme methods for achieving their goals may normalize or glorify violence. As a result of repeated exposure, violence can increasingly seem effective, legitimate, normal, and even a sublime form of human expression. This denies the potential of violence to damage relationships, people, the environment, and the constructiveness of the conflict resolution process. Although within-group ideological diversity that includes extremists can be difficult and distasteful, slow progress, and seem inefficient in the short run, in the long run its inclusiveness can promote environmentalism by forging creative, integrative, far-reaching, and durable solutions to environmental conflicts.

Denial of Self-Involvement

Although we tend to think of moral exclusion as excluding others, moral exclusion involves self-exclusion as well. *Involvement* means willingness to take action, to allocate resources, to be concerned about others, and to make sacrifices that ameliorate an environmental problem. It is therefore consistent with moral inclusion and environmentalism. As Table 2 indicates, denial of self-involvement is evident in such symptoms of moral exclusion as displacing blame for harms on those harmed; believing that one's contribution to an environmental problem is undetectable; denying personal responsibility for environmental harm by seeing it as the result of collective rather than individual decisions and actions; and casting oneself as a clean and blameless outsider in comparison to dirty, irresponsible, reprehensible stakeholders. Denial of self-involvement takes two forms: self-exclusion and reluctant participation.

Self-exclusion. Denial of self-involvement minimizes the extent to which an environmental dispute is relevant to oneself or one's group. Individuals, groups, or polities seen as stakeholders to the conflict by some may hope to exclude themselves as affected by the problem or hope to exclude an environmental issue from the scope of their concerns in order to protect their self-interest. By casting themselves as "clean" and insignificant contributors to pollution, they assert their nonrelevance to environmental controversy. This is exemplified in individuals who do not see their part in contributing to air pollution when they drive alone to work each day or purchase sport utility vehicles emitting high levels of emissions.

We also self-exclude to protect our sense of well-being. Although we would like to see ourselves as safe, protected, and able to assess our vulnerability to environmental harm with some certainty, this is not realistic. Environmental conditions depend on natural and human systems and are subject to rapid, unpredictable change. Environmental conditions also result from long-term, cumulative harms that take years, decades, or longer to surface (Susskind & Cruikshank, 1987). Therefore, identifying who is likely to be affected by air, water, or soil pollution, including ourselves, is not always possible.

Reluctant participation. Some stakeholders reluctantly participate in environmental conflict as negotiators but they do so for pragmatic reasons: to comply with federal mandates, to protect the interests of their region, or to prevent the adoption of an agreement they or their constituency would find onerous. More engaged stakeholders, such as downwind states affected by transported air pollution from upwind states, view the reluctant participation self-proclaimed outsiders as denial. During ozone transport negotiations, some states sought to exclude themselves from having to implement controls by denying that they were contributing to the ozone transport problem. Instead they blamed downwind states: "Utilities in non-attainment areas facing draconian controls are looking for the strictest possible ozone controls on the rest of the OTAG region to help ease their attainment burden. But utilities in attainment areas are resisting this" ("Ozone transport region utilities," 1996, p. 7). Other states that were unwilling to assume responsibility and engage in action claimed special privileges and denied their regions' contribution to pollution.

Another route to denial of self-involvement recognizes a problem as real but focuses on parts of the problem that exonerate oneself and therefore denies one's contribution to the problem. Although an upwind region may agree that interstate ozone transport poses public health concerns, it may identify the source of the problem as another region rather than acknowledge its own contribution. When Governor George Voinovich of Ohio stated that "the Northeast creates 75 percent of the problem it is now pointing to the Midwest to solve . . . Ohio contributes less than 5 percent of the smog problem in the Northeast" (Voinovich, 1997, p. A19), he acknowledged the issue but denied Ohio's contribution to it, locating the predominant source of smog in downwind states.

In sum, denial resolves the inherent complexity and ambiguity of environmental conflict by simplifying facts and issues and by replacing uncertainty and the unknown with dogma. In so doing it blunts the challenge and impetus for social change inherent in these conflicts. As a result, trust, communication, cooperation, and constructive conflict resolution are casualties of the conflict.

Conclusion

Environmental controversies are complex, multiparty disputes. Although the physical facts, stakeholders, and their interests differ considerably from conflict to conflict, the processes that underlie environmental conflicts have striking similarities. Fundamental justice beliefs, underlying moral issues, and denial shape the course of environmental conflicts and influence the analysis of "facts" in the controversy, the allocation of blame, the assessment of one's own contributions to the issue, and the evaluation of trade-offs that can resolve the conflict. The metaphor of cancer is apt (cf. Sontag, 1978). Denial is undetectable at very early stages, but as it gains energy, its spread is insidious and it actively attacks its host. Although denial and moral exclusion are more obvious in escalated conflict, it is particularly important to recognize denial and moral exclusion in chronic, nonescalated conflicts. These are the conflicts that are most common and most amenable to conflict resolution.

Conflict escalation is itself a form of denial; salient issues are lost as conflicts shift in focus and spiral outward. Conflict volatility and expansion are dramatic and can distract from—and be preferable to—addressing deeper issues in a conflict, such as the seemingly insoluable, volatile identity issues embedded in protracted intergroup conflicts (cf. Rouhana, 1997). Ironically, the intensification of environmental conflict and degradation can be positive when it makes denial of outcome severity, stakeholder exclusion, and self-involvement more difficult. As a consequence of conflict intensification, negative environmental and human effects are increasingly obvious, the urgency of finding long-term solutions increases, and public support for scrutiny and regulation of previously acceptable practices increases. Thus, environmental disbenefits have the potential to activate the individual and collective concerns that can increase the scope of justice.

Implications for Theory

Utilizing environmental conflicts as a vehicle, this article has identified three kinds of denial in environmental conflict. Whereas previous research has addressed the antecedents (Opotow, 1987, 1993) and consequences (Opotow, 1994) of moral exclusion, this article has focused on its process. The three kinds of denial we have identified conceptually differentiate among the symptoms of moral exclusion and bring some order to conflict processes as they unfold in real time. Our analysis suggests that denial is not merely reactive and the result of conflict but is instead powerfully proactive. The three kinds of denial we identify function as interdependent gears that drive the process of moral exclusion. They do so by aggressively changing perceptions, motivations, morals, and behavior, facilitating the tunnel vision that minimizes complexity of the issues, facts, and parties.

Implications for Practice

Given that denial is part of environmental conflict, how can the process be managed constructively? Ozawa and Susskind (1995) identify three techniques that lend themselves well to constructive management of environmental conflicts:

information sharing, joint fact finding, and collaborative model building. Employing these techniques can bring policymakers and other stakeholders to the table as direct participants with scientists, resulting in more grounded decision making and enhancing communication, perspective taking, and credibility.

Our analysis suggests that the focus should also be on identifying and managing denial, maintaining the integrity of a conflict resolution process, and fostering constructive process and outcome. A constructive conflict resolution process includes the following elements:

- *Transparent processes:* Are they open to inspection and negotiation?
- *Interdependencies:* Are they understood and valued?
- *Perspective taking:* Is it occurring early and often?
- *Inclusion and access:* Are stakeholders' concerns and perspectives at table?
- *Looking ahead:* Are the changes that can occur over time being considered?

Although denial thwarts solutions to environmental problems, collective involvement and the inclusion of diverse stakeholders facilitates communication, perceptions of interdependence, trust, and collective problem solving, offering processes conducive to lasting and constructive solutions. Perspective taking is crucial. Like lungs breathing in and out, perspective taking is a shifting from background to foreground. Without a rhythm and regularity to this shift, much is lost.

A final form of denial bears mention. Seeing the environment as "out there" or as "other" instead of within ourselves exhibits exclusion and denial. The natural world is internal. It is the air we breathe, the water we drink, and the elements from which our bodies are constituted and continuously remade. Seeing the environment as separate from oneself creates a false distinction coloring our sense of the interdependencies between self and environment.

We close with three simple tenets. Keeping them in mind can thwart the tendency for denial and moral exclusion in environmental conflict. First, we are all victims in that we are recipients of pollution generated by others. Second, we are all violators in that we create pollution that has an impact on others. Third, we all need to work at ongoing, constructive problem solving and dialogue. Acting on these tenets takes persistence but it can minimize environmental damage and foster environmental benefits for all.

References

Clayton, S., & Opotow, S. (1994). Green justice: Conceptions of fairness and the natural world. *Journal of Social Issues, 50*(3).

Corsini, R. J. (1999). *The dictionary of psychology.* Philadelphia: Bruner/Mazel.

Deutsch, M. (1973). *The resolution of conflict.* New Haven, CT: Yale University.

Deutsch, M. (1985). *Distributive justice.* New Haven, CT: Yale University.

Hardin, G. (1968). Tragedy of the common. *Science, 162,* 1243–1248.

Jacoby, R. (1975). *Social amnesia.* Boston: Beacon Press.

Krebs, D., Denton, K., & Higgins, N. C. (1988). On the evolution of self-knowledge and self-deception. In K. B. MacDonald (Ed.), *Sociobiological perspectives on human development* (pp. 103–139). New York: Springer-Verlag.

Leopold, A. (1949). *A Sand County almanac.* New York: Oxford University Press.

Lovelock, J. E. (1979). *Gaia: A new look at life on Earth.* Oxford: Oxford University Press.

Mather, L., & Yngvesson, B. (1980–81). Language, audience, and the transformation of disputes. *Law & Society Review, 15*(3–4), 775–821.

Merchant, C. (1980). *The death of nature.* San Francisco: Harper & Row.

Mullen, J. D. (1995). *Kierkegaard's philosophy: Self-deception and cowardice in the present age.* Lanham, MD: University Press of America.

Opotow, S. (1987). Limits of fairness: An experimental examination of antecedents of the scope of justice (Doctoral dissertation, Columbia University, 1987), *Dissertation Abstracts International, 48,* B2500.

Opotow, S. (1990). Moral exclusion and injustice: An introduction. *Journal of Social Issues, 46*(1), 1–20.

Opotow, S. (1993). Animals and the scope of justice. *Journal of Social Issues, 49*(1), 71–85.

Opotow, S. (1994). Predicting protection: Scope of justice and the natural world. *Journal of Social Issues, 50*(3), 49–63.

Opotow, S., Weiss, L., Lemler, J., & Brown, T. (1997). *Air, sea, and land: Environmental conflicts and the scope of justice.* Paper presented at the annual August meeting of the American Psychological Association, Chicago.

OTAG result: Severe Nox emission cuts for midwest utilities; less for mobile sources. (1996, September 12). *Inside OTAG, 1*(2), p. 8.

Ozawa, C., & Susskind, L. (1995). Mediating science-intensive policy disputes. *Journal of Policy Analysis and Management, 5*(1), 23–39.

Ozone transport region utilities quietly draft moderate position. (1996, September 12). *Inside OTAG, 1*(2), p. 7.

Rouhana, N. N. (1997). *Palestinian citizens in an ethnic Jewish state: Identities in conflict.* New Haven, CT: Yale University Press.

Sontag, S. (1978). *Illness as metaphor.* New York: Farrar, Straus and Giroux.

Stern, P. C., & Dietz, T. (1994). The value basis of environmental concern. *Journal of Social Issues, 50*(3), 65–84.

Stone, C. D. (1974). *Should trees have standing? Toward legal rights for natural objects.* Los Altos, CA: William Kaufmann.

Susskind, L. (1981). Environmental mediation and the accountability problem. *Vermont Law Review, 6*(1), 1–47.

Susskind, L., & Cruikshank, J. (1987). *Breaking the impasse: Consensual approaches to resolving public disputes.* New York: Basic Books.

Susskind, L., & Field, P. (1996). *Dealing with an angry public: Mutual gains approach to resolving disputes.* New York: Free Press.

Thompson, L. L., & Gonzalez, R. (1997). Environmental disputes: Competition for scarce resources and clashing values. In M. H. Bazerman, D. M. Messick, A. E. Tenbrunsel, and K. A. Wade-Benzoni (Eds.), *Environment, ethics, and behavior* (pp. 75–104). San Francisco: New Lexington Press.

Voinovich, G. V. (1997, August 12). The clean-air war. *The New York Times,* p. A19.

Weiss, L. (1996). *Justice issues and the negotiations of the Ozone Transport Assessment Group (OTAG).* Unpublished manuscript, University of Massachusetts Boston.

Weiss, L. (1997). *Denial and canonization: Anne Frank's diary.* Unpublished manuscript, University of Massachusetts Boston.

SUSAN OPOTOW is an Associate Professor in the Graduate Program in Dispute Resolution at University of Massachusetts Boston. Her research examines exclusion from the scope of justice theoretically and its applications to environmental, school, and public policy issues. She was editor of a 1990 *Journal of Social Issues* issue on "Moral Exclusion and Injustice," a 1992 *Social Justice Research* issue on "Affirmative Action and Social Justice," and coeditor (with Susan Clayton) of a 1994 *Journal of Social Issues* issue on "Green Justice: Conceptions of Fairness and the Natural World."

LEAH WEISS lives in Annapolis, Maryland. She has 15 years of experience in the environmental regulatory field. She holds a master's degree in Urban and Environmental Policy and Civil Engineering/Public Health from Tufts University and a certificate from the Graduate Program in Dispute Resolution at the University of Massachusetts at Boston. Her exploration of denial and conflict is rooted in her interest in conflict theory and her professional involvement in multiparty environmental negotiations.

Journal of Social Issues, Vol. 56, No. 3, 2000, pp. 491–508

Human Nature and Environmentally Responsible Behavior

Stephen Kaplan*

University of Michigan

This article constitutes a search for a people-oriented approach to encouraging environmentally responsible behavior. It attempts to provide a source of motivations, reduce the corrosive sense of helplessness, and generate solutions to environmental problems that do not undermine the quality of life of the people who are affected. The altruism-centered approach currently popular in the academic literature, by contrast, is seen as contributing to helplessness and focusing on sacrifice rather than quality-of-life-enhancing solutions. An alternative, the Reasonable Person Model, offers an evolutionary/cognitive/motivational approach to understanding human nature.

Facilitating the adoption of environmentally responsible behavior (ERB) is a major challenge for the behavioral sciences. As with any problem, how one approaches it and whether or not it can be solved depend to a large degree on how the problem is conceptualized (Bardwell, 1989; Posner, 1973). In the research literature a prominent approach to this difficult issue has cast the problem as essentially motivational, focusing on altruism as a crucial motive to study (De Young, this issue). The altruism-centered approach is seen as having several inadvertent consequences, including contributing to helplessness and stressing sacrifice rather than quality-of-life-enhancing solutions.

The purpose of this article is to propose an alternative approach that avoids some of the limitations inherent in the altruism-centered approach. This alternative approach has three goals: to provide a durable source of motivation, to reduce the corrosive sense of helplessness, and to generate innovative solutions that people do

---

*Correspondence concerning this article should be addressed to Stephen Kaplan, Professor, Department of Psychology, University of Michigan, Ann Arbor, MI 48109-1109 [e-mail: skap@ umich.edu].

not perceive as threatening their quality of life. Achieving these goals requires more than ad hoc proposals or patchwork solutions. Rather, the approach must be based on a coherent conception of human nature that speaks to the relationship between how people approach new information, how information relates to motivation, and how information and motivation relate to behavior change. Although such a conception may well be beyond what is currently possible, the Reasonable Person Model is proposed as a first approximation to such a conception of human nature. Before describing this framework and proposing the alternative approach, I begin by explaining my reservations about altruism and explore some hypotheses concerning why altruism is so popular and why this popularity may be ill-placed.

Altruism in Perspective

Although efforts to encourage ERB have called upon a wide range of motivations, a substantial portion of the scholarly literature on this topic has focused on altruism. Informal observation suggests that the tendency to focus on altruism is also characteristic among students concerned about environmental problems. In their view there is an inherent linkage between "good" motives and "good" behavior. In addition to the implicit moral issue (i.e., behavior motivated by altruism is seen as morally superior), there appears to be an assumption that there should be a symmetry between the moral value of the motive and the moral value of the action.

One way to cast some light on this widely held assumption that there is a close relationship between the goodness of a motive and the goodness of the behavior that it motivates is to see how difficult it is to find counterexamples. There are three distinct ways to approach such a search. One could look at psychological analyses of evil behavior to determine if the antecedent motivations are commensurately "bad." Alternatively, one could reverse the order, looking at instances of "good" motives in search of instances in which the resulting behavior could be considered "bad." Finally one could search for instances in which less well regarded motives have led to desirable outcomes. Let us examine each in turn.

Some Failures of the "Good Motives Lead to Good Behavior" Assumption

Becker's (1975) *Escape From Evil* provides a potent introduction to the topic of antecedents of bad outcomes. It is a brilliant and moving little book that takes on the challenge of explaining the existence of evil on a large, societal scale. Taking the Holocaust as his paradigmatic example, Becker bases his explanation on three interacting motives. The first is the need to belong to a group, and the second is the need to rise above the group. Given the ubiquity and innocence of these familiar human motivations, one looks expectantly toward the third motive for something more powerful and more unacceptable. His third proposed motive thus comes as something of a surprise; it is to stamp out evil. In the context of Becker's examples it becomes clear that chaos or disorder serve as prime examples of the evil that people feel a need to stamp out. Taken as a whole, these three motives could as well describe the efforts of any of us who have tried to have our clarifying insights both accepted by the group and recognized as memorable accomplishments.

The question of whether good things follow from well-intended motives can be viewed from several perspectives. Among the motives widely acknowledged as "good," altruism must be considered one of the most frequently mentioned. As it turns out, counterexamples of the assumed linkage between altruism and positive behavior are by no means difficult to find. Koestler (1970) identified loyalty to a cause as one of the most distinctive capabilities of the human species—and also as cause of more suffering than any other single factor. Baumeister's (1997) analysis of a vast empirical literature in *Evil: Inside Human Cruelty and Violence* offers a conclusion that closely parallels Koestler's analysis. Baumeister identifies idealism as one of the primary causes of cruelty and violence.

Helping others for its own sake also falls under the larger heading of altruism. Langer (1989), among others, has commented on the great, albeit unintentional damage done by such help, as when an elderly individual receives unneeded help, subtly fostering helplessness. At a quite different scale, Scott (1998) has documented heroic efforts to help humanity at the level of the state. He describes revolutionary efforts to improve the human condition through what were believed to be scientific principles. Among his examples are the institution of collective farming in the Soviet Union. Pol Pot's Cambodia and Mao Tse-tung's China provide other chilling examples.

Holmes (1990, p. 271) provides a trenchant summary of arguments against the idea that selfless motives guarantee desirable outcomes: "If our concepts precommit us to the idea that whenever people overcome selfishness they necessarily act in a morally admirable matter, then our vision of political behavior is bound to be blurred. A moralized selfish/selfless scheme blinds in precisely this manner. Thus, any vocabulary lacking the category 'selfless cruelty' is historically impoverished. It also provides an inadequate guide to the thought of the keenest observers of early modern Europe."

The third type of search in the examination of the assumed motivation-behavior symmetry concerns "bad" motives leading to "good" behavior. Although it is difficult to define a "bad" motive, there are clearly certain motives widely regarded as less desirable. One of these is expressed by the NIMBY (not in my backyard) phenomenon. People unhappy with locally undesirable land uses (LULUs) located near them are often considered to be motivated by selfishness and are seen as quite willing for the same LULUs to be located in someone else's backyard. Yet the vast grassroots movement concerning the siting and management of toxic waste facilities started in the backyards of people living near the infamous Love Canal. Lois Marie Gibbs wrote of becoming transformed by this experience from being a housewife to being a leader of the first citizens' organization to fight hazardous waste in the Foreword to *Not in Our Backyards* (Freudenberg, 1984), the book that chronicled the national movement that grew out of this initial organizing effort.

Another demonstration of the positive power of less well regarded motives occurred in the aftermath of the Detroit race riots of 1943. The discovery that the Detroit police actively participated in the riots prompted the police force to bring in psychologists to teach tolerance. This intervention, however, was found to have no effect on the behavior of the police. Subsequently a different strategy was explored. Police were informed that any of them found practicing racial discrimination would be fired. This alternative approach is reported to have been remarkably successful.

Some people regard the acquisitive motive, leading to the amassing of large amounts of money and property, as undesirable. Nonetheless, such acquisitiveness has in a number of instances resulted in the preservation of valuable natural landscapes. A final example involves the provision of window boxes to residents of an inner-city neighborhood (Lewis, 1996). Pride, a sometimes-shunned motive, appears in this case to have led to a widespread cleanup effort and a resulting increase in the quality of life of the residents.

Altruism in the Environmental Context

Most of the evidence undermining the assumed linkage between alternative motives and "good" behavior is not directly related to ERB. Does this mean that the altruistic focus might be harmless—or even beneficial—in this context? As it turns out this is not the case; altruism has a serious limitation in this role. The reason for this limitation is embedded in the peculiarities arising from the traditional definition of altruism, which is strikingly different from the way other motives are defined. Altruism is defined as feeling or acting on behalf of the welfare of others in cases where self-interest could not be involved (Jencks, 1990). Thus, to the extent that altruistic action involves any cost or effort, it necessarily entails sacrifice, since there cannot be a compensating benefit to the self. In fact, research in this area typically focuses on "situations in which [people] are demonstrably acting against their self-interest" (Mansbridge, 1990, p. 133) and in several studies, the willingness to sacrifice is what defines ERB. The centrality of such a negative payoff as essential to the definition of altruism creates serious motivational issues that bring into question the strategic usefulness of the concept.

Must virtue be unpleasant? The requirement of receiving no benefit from one's action and the inclination to enshrine sacrifice as a paradigmatic environmental virtue (for example, Gigliotti, 1992) communicate a powerful, if unintended, message, namely that ERB inherently leads to a reduction in the quality of life. As De Young (1990–91, p. 216) points out, "While frugality may be accepted as a necessary feature of the future it is usually portrayed as an onerous undertaking, one requiring personal sacrifice of the highest order. People, it is argued, are being asked to give up a modern, high-technology existence for an austere, bleak but needed substitute."

This "dour environmentalist" stereotype to which the altruism concept unintentionally contributes is far from helpful. The hope for a better future is a characteristic of the human makeup that has been found across a wide range of cultures (Cantril, 1966). Casting a negative pall on this hope is unlikely to be an effective motivational strategy. In addition to this unfortunate side effect, the focus on sacrifice embodies a key assumption that may not be well grounded. It assumes that we know what has to be given up. This creates an image of future deprivation that may be unnecessary and inaccurate. Asking people how willing they are to give up x implies that x will be one of the activities that will no longer fit in an environmentally correct future. The impression created is particularly negative because the "giving up" is not placed in the context of environmentally preferable alternatives.

Thus the focus on altruism brings with it the implicit message that living with less will result in an impoverished and joyless future. Ironically, this perspective buys into the establishment view that the way we do things now is the most satisfying, that there is a positive relationship between resource use and happiness, and that materialism and waste are more fun. Perhaps not surprisingly, this presumed linkage between consumption and happiness does not stand up to a careful psychological analysis (Myers & Diener, 1995).

Not only does the dour, negative view of the future put the worst possible face on appeals for ERB, it also undermines potentially powerful linkages with critiques of materialism as unhealthy and unsatisfying as well as environmentally unsustainable. These critiques offer the opportunity to reverse the "sacrifice" perspective, thus making ERB not a regressive activity, but a possible route to a better life. There are some potentially exciting alliances here. Miller's (1995) *How to Want What You Have* is a thoughtful psychological analysis, blending an Eastern perspective with the rich experience of an insightful clinical psychologist. The "use less stuff" approach to happiness is powerfully presented in Dominguez and Robin's (1992) *Your Money or Your Life*. And the general issue of lifestyles that are at the same time less destructive and more satisfying is addressed in Johnson's (1985) *The Future Is Not What It Used to Be*.

Motivational reality. Are appeals to sacrifice, to behaving counter to one's self-interest, a realistic approach to motivating behavior? Is such purity an appropriate characterization of people deeply committed to ERB? Mansbridge (1990, p. 133) puts the issue succinctly: "We normally see self-interest and altruism as being at opposite poles. Indeed, conceptually we know what we mean by altruism only by contrasting it with self-interest. In practice, however, altruism must coincide with self-interest sufficiently to prevent the extinction of either the altruistic motivation or the altruist."

But is such a dichotomy necessary? Does it, in fact, characterize committed environmentalists, past and present? Henry David Thoreau is regarded by many as one of the great environmentalists of all time. However, he fails the test of altruism: He obtained great satisfaction from nature and did not wish to see his country adopt materialism, for his own sake as well as theirs. Most environmentalists of my acquaintance are no different. They love nature and they treasure the benefits they experience from it. They fear the impact its destruction would have, for their own lives and the lives of their descendants, as well as for humanity as a whole.

Research a colleague and I carried out in the context of a wilderness program provides further support for the coexistence of the strong concern for preserving the natural environment and the desire to have such settings available for one's own joy and peace of mind (R. Kaplan & S. Kaplan, 1989). Program participants were asked to keep diaries for the first week after their return from the 10-day program. The powerful and far-reaching effects of the program were evident in these reflections as participants wrote of their feelings of "wholeness" and "oneness"; they considered the wilderness experience to have been transforming, to have revealed aspects of their personalities that they had not known before and that they now valued highly. Their clear concern with protection of the natural environment was deeply personal and closely tied to their hopes for the availability of such transforming experiences in the future. At the same time, however, their concern for preserving the resource closely matched Gagnon Thompson and Barton's (1994) discussion of "ecocentric" values as stemming from "a spiritual dimension and intrinsic value in [the] experiences in nature and feelings about natural settings" (p. 150).

An Alternative Approach

It may be helpful to take stock of the nature of the beast and the nature of the problems it faces.

Reconceptualizing Human Nature: The Reasonable Person Model

At an intuitive level it is not surprising that people resist making changes that they perceive as reducing their quality of life. It is also not surprising that people are concerned about the future of the environment. Perhaps a broader view of human nature, one that encompasses more than material gain, could provide a way out of this impasse.

A central failing of the altruistic position is that it attempts to put aside the issue of gain, of self-interest, in human behavior. The "economic man" position, by contrast, argues that gain is all that matters. Neither position is satisfactory; there is need for a position that is neither so extreme with respect to the issue of gain nor so narrow in its focus.

Even though people are often not rational in terms of the formal (economic man) meaning of the concept, they are clearly capable of being reasonable in the sense of behaving as one would hope people would behave. At the same time it is abundantly evident that people have an enormous capacity to be unreasonable as well. And the very same individual can be reasonable at some times and unreasonable at others. This suggests that it is the circumstances in which people find themselves that may well play a central role in their behavior.

What is it about the circumstances in which people find themselves that could make so radical a difference? Some useful insights can be found in two largely unrelated fields, namely cognitive science and human evolution. The interdisciplinary field of cognitive science has made many important contributions. One of these, perhaps too readily overlooked, is the repeated demonstration that people are exceptionally adept at processing information. In many information-processing areas, such as language understanding and object recognition, human competence still vastly exceeds the capacity of high-speed computers, despite the large investments of time and money over many years.

According to students of evolution this remarkable capability is no accident. Rather, it is due to the niche that humans occupied at a critical point in the evolutionary process (Berrill, 1955; Lachman & Lachman, 1979; Laughlin, 1968; Pfeiffer,1972; and Washburn, 1972). When our primate ancestors came down from the trees to live in a savanna environment, they found a terrain already inhabited by well-adapted competitors. Although no comfortable niche was uninhabited, quick and skillful information processing made it possible to survive on the margin, utilizing foresight and flexibility, supplemented with considerable knowledge of the environment, of potential predators, and of possible prey. If indeed early humans depended for their very survival on information processing, they would be expected to have strong feelings about it. Finding oneself in a confusing environment, for example, could not have been a situation to take lightly. Those early humans who survived must have had an inclination to behave quite differently in circumstances supportive of their information-processing capability as opposed to circumstances that made that activity difficult or impossible.

In this perspective, humans can be seen as active, curious, problem-solving animals. Given survival pressures, processing information had to occur with speed and solving problems with ingenuity. Further, early humans who took delight in these adaptive activities had a decided advantage, since they would have been inclined to practice them in their free time, an activity sometimes referred to as "play." It would also have been adaptive to avoid environments and situations in which such activities were difficult or ineffective. Here again motivation played an important role. Early humans with a strong distaste for such environments would have had an adaptive advantage.

The Reasonable Person Model draws on these cognitive and affective themes. By recognizing human inclinations and the circumstances that are supportive of human motivations, it may be easier to get people to behave in environmentally responsible ways without calling on guilt or sacrifice. As a first, rough step in that direction, I have identified, with a colleague, three aspects of information processing that on evolutionary grounds would be expected to have strong behavioral and motivational implications (S. Kaplan & R. Kaplan, 1989):

- People are motivated to know, to understand what is going on; they hate being confused or disoriented.
- People also are motivated to learn, to discover, to explore; they prefer acquiring information at their own pace and in answer to their own questions.
- People want to participate, to play a role, in what is going on around them; they hate being incompetent or helpless.

The final principle is particularly important in the present context. A situation in which people cannot act effectively, in which they cannot solve the problems they face or cannot implement the solutions they come up with, is likely to be extremely distasteful. In other words, people would be expected to avoid contexts that they consider conducive to helplessness. And since this is a cognitive animal, one would expect an avoidance of even thinking about realms that evoke feelings of helplessness. Thus, in this perspective, helplessness would be one of the most important motivational issues to consider in the context of behavior change.

Research suggests that helplessness is not only an important issue in the context of ERB, it is perhaps *the* pivotal issue. Levin (1993) reports a cross-cultural study of reaction to the increasing quantity of information available concerning environmental problems. The study, by London-based Research International, found that more information led not only to greater concern but also a greater sense of helplessness. More recently, a study by the nonprofit group Public Agenda found substantial declines in concern about environmental issues. This decline is attributed not to apathy but to a sense of futility and helplessness (Donn, 1999). This suggests that many who appear uninterested in environmental issues may distance themselves to avoid pain, not because environmental issues are of no concern to them. Although people may vary in what they consider to be an "environmental" issue, few favor sprawl, few prefer polluted air, or unsafe water, or decline in fish stocks, or news of people starving because of population explosion and environmental disaster.

Additional support for the corrosive effect of helplessness on ERB comes from a surprising source. A recent test of Geller's "actively caring" hypothesis by Allen and Ferrand (1999) looked at the impact on self-reported ERB of three of Geller's factors predicting actively caring (self-esteem, personal control, and belonging). Only one of these predictor variables, personal control, showed a significant relationship. In relating this research to the present context it is essential to take a close look at Allen and Ferrand's measurement of personal control. The authors state that this construct "was assessed in terms of the extent to which participants felt their action could benefit the environment" (p. 342). This is not control in any usual sense of the word. A far more accurate characterization would be "the opposite of helplessness." Thus the core finding of the study could be described as showing that people who feel helpless, who feel that their behavior would not make a difference, are less likely to participate in ERB. Taking this result along with the striking parallel findings of the two surveys leads one to conclude that any psychological approach to ERB that does not directly address the helplessness issue may have limited practical value.

Elements of Solutions

If altruism and sacrifice are unlikely to achieve the desired results, we need to consider the goals for a viable alternative. These would seem to entail finding ways to motivate people to be environmentally responsible in a way that also reduces their sense of helplessness and, at the same time, is sensitive to their needs and inclinations.

The general format of the proposed solution is participatory problem solving. "Participatory" implies that many people could be engaged in such activity. The expression "problem solving" constitutes a reminder that the purpose of the participatory activity is not to implement plans that someone else has already drawn up but to find innovative solutions to environmental problems while meeting the needs of the participants.

The elements of solutions offered here have all been used in many situations, as some of the examples testify. Despite their relatively widespread use, however, they have not been examined in terms of their motivational implications. This is particularly true of their relationship to the principles of the Reasonable Person Model. These principles play an important role in understanding the effectiveness of the various elements of the solution discussed here.

A focused task that needs problem solving. Rather than telling people what they must do or do without, the proposed approach provides people with an opportunity to figure out for themselves how various broadly defined goals can be met. In order to achieve this, they must have a specific focus, a particular problem that they are trying to solve. Such task-oriented groups could arise in many contexts: in business, in government at many levels, in various agencies and organizations, and as grassroots efforts.

Even at a larger scale of federal decision making, such an approach is not without precedent. The Netherlands has undertaken to achieve sustainability in one generation. This "Dutch Green Plan" is based on a public-private partnership that is particularly striking for the way it allocates responsibility. The government identifies environmental problems according to themes, and then sets targets and objectives within each. Target groups are then identified (such as industry or consumers) that are responsible for achieving these objectives. In this way innovation is encouraged and participation is widespread (AtKisson, 1995).

The challenge to find a solution that is both satisfying and responsible. Whereas altruism calls upon the heroic (or perhaps the saintly), the proposed approach looks to what is reasonable. Counting on prolonged and generalized sacrifice may work in some circumstances (William James [1910/1971] struggled to discover the "moral equivalent of war"), but the current challenge is unlikely to be one of them. People are generally more comfortable with activities that represent a confluence of self-interest and altruism than activities that pit one against other (Mansbridge, 1990). Thus the emphasis here is on both the sustainable and the satisfying.

If one is to avoid asking people to make choices that run counter to their perceived self-interest, choices must be available that are not only good for the environment but multiply desirable. Thus one of the goals of the proposed approach is to generate what one might call "multiply desirable choices."

A major benefit of the focus on multiply desirable choices is the avoidance of situations in which people, feeling guilty about their resistance to adopting alternatives they feel are unacceptable, resolve the conflict by tuning out the message and avoiding such messages in the future. It is for similar reasons that Roberts and Bacon (1997, p. 89) argue for alternatives to the individual behavior change strategy: "Other strategies outside the individual awareness and action should be considered. Environmental campaigns must avoid 'blaming the victim' strategies. Individual behavior change strategies are inappropriate if macro conditions exist which can be blamed for contributing to the problem or constraining the effectiveness of individual efforts (e.g., companies that do not provide ecologically friendly products, government inactivity)."

An interesting example of such an "other strategy" is provided by an alternative to single-occupancy-vehicle trips, a shuttle service, in Boulder, Colorado. This shuttle, called SKIP, was designed with considerable input from a citizen advisory board, which made sure that the plan was sensitive to the need for "friendly drivers, comfortable seats, and a cozy interior to create a safe and comfortable public space" (Renew America, 1998). People using this environmentally preferable alternative (projected to reach 1.47 million annually) may well feel good about the virtue of their actions, but they are not suffering in the process and in fact are likely to have improved their quality of life in the process. With the help of group problem solving, a multiply desirable choice has been created.

Effective participatory problem solving. On the face of it, it might seem unrealistic to put decisions in the hands of the very same people who are currently behaving so inappropriately with respect to environmental issues. This concern is based on an assumption that, although completely understandable, does not stand up to closer examination. To understand the issues involved, it may be helpful to consider the differences among the following: (1) telling people what to do, (2) asking people what they want to do, and (3) helping people understand the issues and inviting them to explore possible solutions. The first is the procedure most often employed. The second, although involving participation in a limited sense, is not the kind of participation envisioned here. The third, which describes the proposed approach, is not participation in the sense of an opinion survey. Rather, it incorporates understanding, exploration, and problem solving as essential components of participation.

Although surveys can yield useful information, it is essential to realize that they often do not tap informed opinions; informed opinions, even on the part of ordinary people, can be very different from uninformed responses. Feinsilber (1994) carried out a "deliberative poll" among a group of 200 citizens concerning the often politically inflammatory issue of term limitations. Before expressing their opinions the group read a neutral position paper and discussed the issues extensively. Although term limitations are generally believed to be highly popular, the results were characterized by moderation and by uncertainty as to whether the benefits outweighed the liabilities. As in Hansen's context, a wide range of information would need to be available to participants in the proposed approach.

It is also important to realize that people in participatory groups prefer to work with experts, rather than on their own (Wandersman, 1979). Thus the appropriately structured group would provide a context of new input, increasing the likelihood of a more measured and thoughtful outcome. In the context of the proposed approach, various kinds of expertise would need to be available. Governmental, corporate, nonprofit, and other groups each have many kinds of expertise to contribute, depending on the nature of the problem. Further, many so-called private citizens have important areas of expertise as well, including knowledge about local conditions and environmentally related ways of life (like farming) as well as information about how similar problems have been solved in places where they might have traveled or lived previously.

A thoughtful context and the availability of a wider range of information can have profound effects on the perspective of participants. An example familiar to members of the academic community is the behavior of student representatives on a faculty committee. As they come to understand the constraints and realities of the situation, they start sounding more and more like the faculty members, much to the dismay of the students they represent.

And finally, the observed environmental irresponsibility of many people cannot be interpreted as a simple example of disinterest or inappropriate attitude or sloth. Often there is a lack of appropriate infrastructure, or of multiply desirable choices, or of cultural support. People have many reasons to resist making sacrifices for the common good, among them the concern that others will cheat, and that they will look like fools. When one adds to this the sense of inadequacy and helplessness as an individual tries to compensate for the inappropriate behavior of huge corporations and governments, it is hardly surprising that the behavior of ordinary people often falls short of being exemplary (Bardwell & Kaplan, 1992).

Perhaps an example would be helpful (Renew America, 1999). Every day for many years thousands of people commuted from New Jersey to New York City by way of a dark, traffic-congested tunnel. This pattern wasted large quantities of fuel and created substantial pollution; it was not environmentally friendly. It should be noted that the 30-to-40-minute drive was not people-friendly either. But for many years there was no multiply desirable choice. Then the Port Authority of New York and New Jersey, in collaboration with a private firm, New York Waterway, reinstituted the ferry service that had in earlier times made the trip from Hoboken to Battery Park. The trip takes a scenic 5 minutes. The ferry is not for automobiles; it serves people and bicycles only. But it runs frequently and connects to transit routes at both ends. Annual ridership is over 2.3 million people. Here behavior changed radically in an environmentally responsible direction. Not surprisingly, people prefer making the environmentally responsible choice when they are not seriously disadvantaged by doing so. Judging people's behavior in the absence of multiply desirable choices has limited usefulness as an indication of these inclinations to behave responsibly.

Thus a reasonably cogent case can be made for the argument that people's participation in a problem-solving context need not reflect their apparently uncaring overt behavior. Nonetheless it would be more reassuring if one could point to concrete examples of such activities that have already taken place.

On first blush, this appears to be a difficult undertaking. There is a literature on this topic, but it is modest and scattered. Examples also appear from time to time in the popular press, but they are hard to access and not described in a consistent way. The community effort that led to the remarkable resurgence of the desolate South Bronx provides a pertinent example (Breslin, 1995). More directly related to ERB, environmental activists and Dow Chemical Corporation in Midland, Michigan, recently collaborated on a far-reaching program that will both reduce the level of toxic wastes and save the company millions of dollars (Feder, 1999). To feel convinced that the proposed approach can work in any reliable way, a more substantial database would be helpful. Fortunately for the last 9 years a far more systematic source of information has become available. Renew America, a nongovernmental organization, has been holding an annual competition for what it calls "success stories." Its Web-searchable database contains the essential information for 1,800 of these success stories. It is now clear that participatory activities with positive environmental impact have been occurring in a wide variety of contexts and on a relatively substantial scale.

Relationship of "Elements of a Solution" to the Reasonable Person Model

Attempts to convince the public of the importance of ERB have often yielded disappointingly modest impact. Many of these efforts have unwittingly also violated the principles of the Reasonable Person Model. For example, eagerness to be convincing can lead to presenting too much information, with the result that the recipients of the information are overwhelmed and their understanding of the issues is compromised. Frequently, the context in which information is presented leaves no opportunity for exploration. A sense of urgency on the part of the generator of the appeal may lead to painting a dire picture that contributes to a sense of hopelessness and helplessness.

The proposed approach addresses each of these issues directly. Understanding is enhanced by providing the opportunity to ask questions, to work with experts in the area, and to study pertinent material. Exploration is basic to the process, since playing with innovative solutions is built into the nature of the task.

The way the proposed approach deals with helplessness is a bit more complicated. Part of the complication is conceptual. There is an inclination in the psychological literature to treat control as the opposite of helplessness. For many reasons, this is an inappropriate contrast (Antonovsky, 1987; S. Kaplan & R. Kaplan, 1982; Little, 1987). A far more appropriate counter to helplessness is participation. Although we all want many things to be *under* control, the effort and responsibility involved in control is often seen as decidedly unattractive. By contrast, people often find genuine participation (as opposed to the pro forma variety) satisfying and empowering (S. Kaplan & R. Kaplan, 1989; Wandersman, 1979). Further, the proposed approach allows people to discover that they are not alone, that there are indeed other concerned people. They also benefit from doing something that could make a difference. And finally, to the extent that the proposed process generates multiply desirable choices, the individual may feel both that it is possible to behave responsibly without worrying about the implications of making a sacrifice in the process and that under these conditions others are more likely to join as well. The realization that other groups in other places are also generating multiply desirable choices may also reduce feelings of helplessness, as fear that nothing can be done is replaced by discovery that a great deal is, in fact, being done.

Research Opportunities

Although there has been growing awareness of the importance of participation in environmental design (Hester, 1996; Wandersman, 1979), environmental justice (Ervin, 1992), and community development (Perkins, Brown, & Taylor, 1996), there has been relatively little recognition of its potential in the context of ERB. This potential is, in fact, considerable; participation has the potential to play multiple roles in fostering ERB. Not only can it strengthen positive motivations; participation can also be a powerful factor in reducing the negative motivation of helplessness. In addition, and closely related to its motivational effects, participation can play a pivotal role in finding innovative solutions to environmental problems. Thus the potential usefulness of participation is inextricably linked to the major reframing proposed here. As we have seen, given how the problem of ERB is frequently framed, the necessary actions are assumed to be known; the problem is only to get people to carry them out, acknowledging that some sacrifices may be necessary for the common good. By contrast, if one acknowledges that there may be solutions that do not necessarily involve sacrifice and may even enhance quality of life, then discovering these solutions is as important as motivating responsible behavior.

The availability of a great many examples, such as those provided by the Renew America (1999) Web-searchable database of success stories, is a source of encouragement and reassurance. But it is also much more than that. In conjunction with the proposed framework, both this material and the implicit link it provides to other ongoing or about-to-start participatory problem-solving efforts lend themselves to a wide variety of research approaches, including small experiments (R. Kaplan, 1996) and comparative case analyses, as well as more traditional research designs.

Research issues concerning motivation. There are several kinds of motivations one might expect to be enhanced by the participatory problem-solving experience. One hypothesized cluster involves engagement, connection, a sense of commitment (Brickman et al., 1982), and a sense of ownership (Kearney & Kaplan, 1997). A second cluster might include clarity enhancement, reduced helplessness or empowerment, having a chance to play a meaningful role (Antonovsky, 1987), and being needed (S. Kaplan, 1990). Any of these motivations could have a powerful influence, both on behavior and on interest in communicating with others about the project and its implications.

Since a downside of the sacrifice perspective is the fear of a reduced quality of life, it would be worth exploring whether participation in a problem-solving-oriented project increases an individual's quality of life, both in the present and as anticipated in the future.

The many dimensions of motivation such as reliability, durability, and generalizability (De Young, this issue) could also be assessed in the context of this sort of research. A related issue worthy of study is the diversity of motivations that might be involved among the individuals participating in any specific project. It would be useful to know if motivations for participating are relatively uniform within a given group, or if some participants are motivated by civic pride, others by competition with other places, or by aesthetics, or being needed, and so on. An indication of this sort of differentiation has been obtained in research on the motivations of volunteers (Clary et al., 1998).

Issues related to the solutions obtained. The proposed approach represents a change in emphasis from "known necessary sacrifices" to discovery of multiply desirable choices, of patterns that are satisfying and responsible. Thus these participatory problem-solving efforts have not only psychological consequences but also concrete products as well. Such concrete products might include programs, opportunities, partnerships, and innovative approaches of all kinds. It might be fruitful to compare these solutions with the old way of doing things. For example, one might expect superior creativity and diversity of solutions. It would also be likely that the degree of local match would be superior. Less clear-cut is the issue of transferability. Would the solutions only apply locally, or would corresponding groups and agencies elsewhere rate the solutions as potentially pertinent to their own settings?

Conclusion

Over the years the passive organism envisioned by the behaviorists has begun to give way to a radically different understanding of the human animal. This animal is seen as attempting to comprehend, to make sense of its world. It is addicted to exploring, to discovering, to finding out. Although it is capable of falling into passivity, it is often at its best, and its happiest, when it feels needed, when it can participate in what is going on around it.

A central purpose of this article has been to consider how one would motivate such an animal to be environmentally responsible, to behave in an ecologically sustainable fashion—or more appropriately given the nature of the organism, how one would assist such an animal in discovering ways of relating to the world that would best promote its welfare and that of its children and grandchildren.

Presumably this would not be accomplished by telling it what to do. There are several reasons to avoid this approach: People are likely to resist doing what they are told to do and may even attempt to undermine the entire effort; furthermore, such an approach would be a waste of talent and ingenuity. Telling people what to do ignores the possibility that there may be significant local variants in how best to achieve a particular goal. Being responsive to such local variation might lead to a diversity of solutions, providing the basis for a culture of exploration, innovation, and involvement that will be both satisfying and responsible.

If telling people what to do is a poor approach, then what? Perhaps these three suggestions will prove more effective:

1. Be sensitive to going with the grain, to recognizing and working with the motivations and inclinations characteristic of this species.

2. Treat the human cognitive capacity as a resource.

3. Engage the powerful motivations for competence, being needed, making a difference, and forging a better life.

These should all serve to counter the pervasive malaise of helplessness. By providing opportunities for understanding, exploration, and participation, effective group problem solving can lead to new multiply desirable choices. In such cases the motivations to behave responsibly and to satisfy self-interested needs are no longer in conflict, but mutually supportive.

References

Allen, J. B., & Ferrand, J. L. (1999). Environmental locus of control, sympathy, and proenvironmental behavior. *Environment and Behavior, 31,* 338–353.

Antonovsky, A. (1987). *Unraveling the mystery of health: How people manage stress and stay well.* San Francisco: Jossey-Bass.

AtKisson, A. (1995). Low country, high hope. *In Context* [On-line], 40. Available: www.context.org

Bardwell, L. V. (1989). *Managing helplessness and enhancing problem definition in the context of undergraduate environment instruction.* Unpublished doctoral dissertation, University of Michigan, Ann Arbor.

Bardwell, L. V., & Kaplan, S. (1992). The impact of the introductory environmental course: Guilt and despair vs. hope and commitment. *Environmental Professional, 14,* 346–350.

Baumeister, R. F. (1997). *Evil: Inside human cruelty and violence.* New York: Freeman.

Becker, E. (1975). *Escape from evil.* New York: Free Press.

Berrill, N. J. (1955). *Man's emerging mind.* New York: Dodd, Mead.

Breslin, P. (1995, April). On these sidewalks of New York, the sun is shining again. *Smithsonian, 26,* 100–113.

Brickman, P., Rabinowitz, V. C., Karuza, Jr., J., Coates, D., Cohen, E., & Kidder, L. (1982). Models of helping and coping. *American Psychologist, 37,* 368–384.

Cantril, H. (1966). *The pattern of human concerns.* New Brunswick, NJ: Rutgers University Press.

Clary, E. G., Snyder, M., Ridge, R. D., Copeland, J., Stukas, A. A., Haugen, J., & Mience, P. (1998). Understanding and assessing the motivations of volunteers: A functional approach. *Journal of Personality and Social Psyhchology, 74,* 1516–1530.

De Young, R. (1990–91). Some psychological aspects of living lightly: Desired lifestyle patterns and conservation behavior. *Journal of Environmental Systems, 20,* 215–227.

Dominguez, J., & Robin, V. (1992). *Your money or your life.* New York: Penguin.

Donn, J. (1999, June 2). Frustration sapping environmental concern, researcher says. Associated Press state and local wire.

Ervin, M. (1992). The toxic doughnut. *The Progressive, 56*(1), 15.

Feder, B. J. (1999, July 18). Chemistry cleans up a factory. *New York Times,* pp. BU 1, 11.

Feinsilber, M. (1994, April 3). Pause for thought. *Ann Arbor News.*

Freudenberg, N. (1984). *Not in our backyards: Community action for health and the environment.* New York: Monthly Review Press.

Gagnon Thompson, S. C., & Barton, M. A. (1994). Ecocentric and anthropocentric attitudes toward the environment. *Journal of Environmental Psychology, 14,* 149–157.

Gigliotti, L. M. (1992). Environmental attitudes: 20 years of change? *Journal of Environmental Education, 24,* 15–26.

Hester, R. (1996). Wanted: Local participation with a view. In J. L. Nasar and B. B. Brown (Eds.), *Public and private places* (pp. 42–52). Edmond, OK: Environmental Design Research Association.

Holmes, S. (1990). The secret history of self-interest. In J. J. Mansbridge (Ed.), *Beyond self-interest* (pp. 267–286). Chicago: University of Chicago Press.

James, W. (1971). The moral equivalent of war. In G. W. Allen (Ed.), *A William James reader.* Boston: Houghton-Mifflin. (Original work published 1910)

Jencks, C. (1990). Varieties of altruism. In J. J. Mansbridge (Ed.), *Beyond self-interest* (pp. 53–67). Chicago: University of Chicago Press.

Johnson, W. (1985). *The future is not what it used to be.* New York: Dodd, Mead.

Kaplan, R. (1996). The small experiment: Achieving more with less. In J. L. Nasar and B. B. Brown (Eds.), *Public and private places* (pp. 170–174). Edmond, OK: Environmental Design Research Association.

Kaplan, R., & Kaplan, S. (1989). *The experience of nature: A psychological perspective.* New York: Cambridge. Republished, 1995, Ann Arbor, MI: Ulrich's.

Kaplan, S. (1990). Being needed, adaptive muddling and human-environment relationships. In R. I. Selby, K. H. Anthony, J. Choi, and B. Orland (Eds.), *Coming of age* (pp. 19–25). Oklahoma City: Environmental Design Research Association.

Kaplan, S., & Kaplan, R. (1982). *Cognition and environment: Functioning in an uncertain world.* New York: Praeger. Republished, 1989, Ann Arbor, MI: Ulrich's.

Kaplan, S., & Kaplan, R. (1989). The visual environment: Public participation in design and planning. *Journal of Social Issues, 45,* 59–86.

Kearney, A. R., & Kaplan, S. (1997). Toward a methodology for the measurement of knowledge structures of ordinary people: The Conceptual Content Cognitive Map (3CM). *Environment and Behavior, 29,* 579–617.

Koestler, A. (1970). The urge to self-destruction. In A. Tiselium & S. Nilsson (Eds.), *The place of value in a world of facts: Proceedings of the Fourteenth Nobel Symposium* (pp. 301–304). New York: Wiley.

Lachman, J. L., & Lachman, R. (1979). Theories of memory organization and human evolution. In C. R. Puff (Ed.), *Memory organization and structure* (pp. 134–190). New York: Academic.

Langer E. J. (1989). *Mindfulness.* Reading, MA: Addison-Wesley.

Laughlin, W. S. (1968). Hunting: An integrating biobehavior system and its evolutionary importance. In R. B. Lee and I. DeVote (Eds.), *Man the hunter* (pp. 304–320). Chicago: Aldine.

Levin, G. (1993, April 12). Too green for their own good. *Advertising Age, 64,* 29.

Lewis, C. A. (1996). *Green nature and human nature.* Urbana, IL: University of Illinois Press.

Little, B. R. (1987). Personality and environment. In D. Stokols and I. Altman (Eds.), *Handbook of environmental psychology* (pp. 205–244). New York: Wiley.

Mansbridge, J. J. (1990). On the relation of altruism and self-interest. In J. J. Mansbridge (Ed.), *Beyond self-interest* (pp. 133–143). Chicago: University of Chicago Press.

Miller, T. (1995). *How to want what you have.* New York: Avon.

Myers, D. G., & Diener, E. (1995). Who is happy? *Psychological Science, 6,* 10–19.

Perkins, D. D., Brown, B. B., & Taylor, R. B. (1996). The ecology of empowerment: Predicting participation in community organizations. *Journal of Social Issues, 52*(1), 85–110.

Pfeiffer, J. E. (1972). *The emergence of man* (2nd ed.). New York: Harper & Row.

Posner, M. I. (1973). *Cognition: An introduction.* Glenview, IL: Scott, Foresman.

Renew America. (1998). The SKIP [On-line]. Available: http://www.crest.org/renew_america

Renew America. (1999). Environmental success index [On-line]. Available: http://www.crest.org/renew_america

Roberts, J. A., & Bacon, D. R. (1997). Exploring the subtle relationships between environmental concern and ecologically conscious consumer behavior. *Journal of Business Research, 40,* 79–89.

Scott, J. C. (1998). *Seeing like a state.* New Haven, CT: Yale University Press.

Wandersman, A. (1979). User participation: A study of type of participation, effects, mediators, and individual differences. *Environment and Behavior, 11,* 185–208.

Washburn, S. L. (1972). Aggressive behavior and human evolution. In G. V. Coelho and E. A. Rubinstein (Eds.), *Social change and human behavior.* Washington, DC: National Institute of Mental Health.

STEPHEN KAPLAN is a Professor of Psychology and of Computer Science and Engineering at the University of Michigan. He has written on numerous topics in environmental psychology, including his most recent book (with Rachel Kaplan and Robert Ryan), *With People in Mind: Design and Management of Everyday Nature* (Island Press). He has chaired or cochaired many doctoral committees in a wide range of fields including psychology, computer science, nursing, natural resources, architecture, and geography. His research takes an evolutionary and environmental approach to such basic processes as perception, cognition, and affect and such applied concerns as participation, expertise, and mental fatigue.

Journal of Social Issues, Vol. 56, No. 3, 2000, pp. 509–526

Expanding and Evaluating Motives for Environmentally Responsible Behavior

Raymond De Young*

University of Michigan

This article contends that while striving to promote environmentally responsible behavior, we have focused attention too narrowly on just two classes of motives. There is a need to expand the range of motives available to practitioners and to provide a framework within which motives can be evaluated for both their immediate and long-term effectiveness. The article then examines a strategy for promoting environmentally responsible behavior that has significant potential. This strategy is based on a particular form of motivation called intrinsic satisfaction. Nine studies are reviewed that have outlined the structure of intrinsic satisfaction. A key theme discussed is the human inclination for competence. This fundamental human concern is shown to have both a general form and a resource-specific version.

Although the search for motives effective at promoting environmentally responsible behavior (ERB) is being enthusiastically pursued, the work so far has been somewhat confined. The vast majority of attention has been given to only two motivations: providing material incentives and disincentives sufficient to make the behavior worth attending to and focusing on the altruistic reasons for engaging in the behavior. There has been relatively little exploration of other, potentially more useful alternatives.

Early attention was given to the use of incentives and disincentives. Scott Geller and his colleagues explored the effectiveness of incentives and disincentives in promoting ERB and established that such behavior can be motivated by the manipulation of material reward, whether token or real (Geller, 1987, 1992; Geller,

---

*This article is based on an address presented at the meeting of the Society for the Psychological Study of Social Issues, Ann Arbor, MI, June 19, 1998. The research discussed in the article was partially funded by a grant from the MacArthur Foundation (96-34311A-WER). Correspondence concerning this article should be addressed to Raymond De Young, School of Natural Resources and Environment, University of Michigan, 430 East University Avenue, Ann Arbor, MI 48109-1115 [e-mail: rdeyoung@umich.edu].

Winett, & Everett, 1982; see also Cone & Hayes, 1980). The last quarter-century has witnessed a continued interest in and expansion of the behaviorist perspective (see, for instance, Geller, 1989). As this article will discuss, however, it was two undesirable properties of this approach that encouraged researchers to pursue other motivations. It turned out that incentives needed constant reintroduction to remain effective and they proved to be less reliable than we had hoped (Katzev & Johnson, 1987).

Altruism is another motive that has received significant research attention. It remains popular among researchers as a powerful, if not the dominant, motive for the adoption of ERB. The major conceptual framework for studying altruism has been the Schwartz moral norm activation model (Schwartz, 1970), although Geller has recently proposed an alternative framework (Allen & Ferrand, 1999; Geller, 1995a, 1995b). Current empirical work identifies both a sociocentric and an ecocentric form of altruism (see Eckersley, 1992; Schultz, this issue). For the concept of altruism to be useful for practitioners, we will need to provide the type of specific guidelines for using altruism that exist for using incentives and disincentives. Unfortunately, altruism may suffer from more than just a lack of procedural guidelines, for as Kaplan (this issue) suggests, altruism may be a fatal remedy (Sieber, 1981).

Environmentally Responsible Behavior as Multiply Determined

There is no scientific reason to narrow the range to just these two categories of motivation. After over a century of psychological research, it would hardly seem necessary to argue in support of the concept of the multiple determination of behavior, but for a variety of reasons single-determination theories remain popular. From an evolutionary perspective, it seems likely that there would be multiple motivations impinging on any given behavior. As philosopher Mary Midgley (1978) has pointed out, human beings want many things, not just one. Furthermore, the many are not reducible to or exchangeable for one. We want clear air, she notes, and clean water. No amount of the one can substitute for a lack of the other. She is troubled by a tendency to seek one central motivation for all that we do, finding such efforts "a misplaced and futile sort of economy."

Empirical evidence has emerged supporting the idea that ERB has multiple antecedents (Schultz, this issue; Stern & Dietz, 1994; Stern, Dietz, Kalof, & Guagnano, 1995; Thompson & Barton, 1994) and that specific behaviors may have distinctly different patterns of initiation (Cook & Berrenberg, 1981; Oskamp et al., 1991). Thus, it seems extremely unlikely that ERB is wholly a function of a single motive and more likely, as Allen and Ferrand (1999) contend, that ERB is multiply determined.

Evaluation Criteria for Behavior Change Techniques

To select from among alternative motives we must determine the conditions under which they are effective. Traditionally the effectiveness of a motive is assessed by predicting the occurrence or frequency of self-reported or observed behavior (see, for instance, Corral-Verdugo, 1997). Alternatively, a motive is shown to be significantly associated with an established measure of environmental attitudes or concern (e.g., the New Environmental Paradigm) in an effort to validate its effectiveness. Such unidimensional evaluation, however, misses the fact that there are many features a motive might possess. These features can be organized into two general categories. Outcome-based evaluations deal with the effectiveness of a technique in isolation, whereas context-based evaluations focus on those factors that moderate the effectiveness of a technique.

Outcome-Based Evaluations

Cone and Hayes (1980) argued in favor of two outcome-based criteria: (a) whether a technique can be reliably implemented by a variety of individuals and (b) its ability to promote durable behavior change (also see De Young, 1993). Clearly, the most straightforward question a practitioner can ask is whether a technique does initiate behavior change. Framed in this way, reliability focuses on the more immediate effects of an intervention and can be measured at two levels. The first level is to assess what proportion of a population is responsive. The second level is to assess whether a technique is still capable of effecting change after repeated presentation to the same individual.

Durability, in contrast, concerns long-term effects. The issue here is whether behavior, once changed, is maintained without repeated intervention by the practitioner. The reliability of a behavior change technique is vital. Yet given the number of environmental problems being faced, we could argue that a vital goal is to create behavior change that is long-term and self-maintaining.

Early on, both reliability and durability emerged as weaknesses of material incentives and disincentives. Numerous researchers reported that although monetary incentives are able to initiate ERB, they seem unable to produce durable behavior change: Behavior returned to baseline levels after the reinforcement was terminated (Dwyer, Leeming, Cobern, Porter, & Jackson, 1993; Katzev & Johnson, 1987). It is unrealistic to require that environmental practitioners perpetually intervene to maintain a single behavior. Their programs, particularly their budgets, rarely allow for such vigilance. Even when an incentive could be partially maintained, by employing an intermittent schedule or token rewards, the results nonetheless prove to be less reliable than hoped. In some studies, participation rates were as low as 8% (Katzev & Johnson, 1987), and, as McClelland and Canter (1981) report, the effects do not last:

The studies indicate that positive financial incentives can lead to some conservation, at least for a limited time (3 to 10 weeks). However, the monies distributed have usually exceeded the value of the energy saved; the effects have often faded over time; and many residents seem unaware of or uninterested in the monies available. (p. 14)

It is now known that reliability and durability can be diminished by a variety of psychological processes. For instance, reduced reliability can result from habituation (Brickman & Campbell, 1971), and motives powerful enough to cause overjustification can reduce durability (see Lepper, 1981; Lepper & Greene, 1978). Both reliability and durability can be diminished by psychological reactance, where the recipient does the opposite of what is demanded (J. W. Brehm, 1966; S. Brehm & J. W. Brehm, 1981). This latter phenomenon is more than just a disturbing theoretical possibility. Reactance effects have been noted in numerous investigations including the study of legal prohibitions (Mazis, 1975) and strongly worded prompts for proenvironmental action (Reich & Robertson, 1979).

The possibility of reactance is not limited to strong coercive techniques. Schwartz and Howard (1981) report a number of situations in which "in the presence of factors most conducive to activating norms favoring helping, decreased rates of helping behavior have sometimes been obtained." The range of possible explanations offered by these authors is revealing: suspiciousness following a high-pressure appeal, psychological reactance, and overjustification when "external pressures to provide aid undermine the internalized motivation to perform altruistic actions."

Finally, reactance is not limited to an intervention's recipient. Evidence is accumulating about the effect on the users of powerful interventions. Even a successful behavior change intervention, one that effectively alters the target behavior, can negatively alter the user's perceptions in two ways: contempt for those people he is influencing and self-contempt. In the former case, the more an intervention restricts the recipient's choice of how to respond to an issue, the more the user of that technique will have a negative perception of the recipient (O'Neal, Kipnis, & Craig, 1994; Rind & Kipnis, 1999). When the intervention does not constrain freedom to think and decide, the user of the technique will have a more positive evaluation of the recipient. Rind and Kipnis (1999) also report that the use of strong intervention techniques results in the user's having significantly lower self-perceptions.

Taken together, these findings suggest that we approach all behavior change situations, even those that appear to have succeeded, with caution. Even with the best of intentions, we can trigger reactance and thus possibly reduce both reliability and durability.

Context-Based Evaluations

Cone and Hayes (1980) also suggested a third criterion that focuses on the context of behavior change. This measure, generalizability, evaluates whether a motivational approach can be effectively applied to other environmental problems, settings, and contexts. This is a long-established concern of research. Another way of conceptualizing this measure is to ask about the generalizability of the effect on a single recipient. Here we are interested in unintended but beneficial side effects, the degree to which a person's adoption of a specific ERB either "spills over" to other settings or promotes the adoption of untargeted but related behaviors (De Young, 1993).

There is theoretical support for the idea that prior behavior is predictive of future behavior (Ajzen, 1991). Usually the prediction is of an identical behavior in a single setting: past household recycling predicting future household recycling, for example. Evidence is emerging that a specific behavior in one setting can generalize to another setting. In a study of office-based conservation programs, it was found that prior experience with general household recycling was effective at predicting general office recycling. Likewise, prior household experience with a particular material, in this instance paper, predicted office conservation behavior with respect to that same material (Lee, De Young, & Marans, 1995), and on-the-job recycling has been reported to carry over to the home (Fusco, 1991). There is also evidence that this effect exists with less specificity. Initial, if limited support, comes from a study of a pilot recycling program in which participation in the recycling effort fostered other conservation behavior (Kreutzwiser, 1991). Perhaps most important is that the fundamental mechanism at work here is likely to be familiarity with a new behavior rather than experience in its direct and literal sense. In a study of the adoption of photovoltaics by utility managers, A. W. Kaplan (1999) reported that conceptual familiarity was an effective predictor of adoption interest. This is an extremely hopeful notion for practitioners, since what people can become familiar with is not limited to what they directly experience.

A related generalizability issue is whether motivational techniques can be designed for universal application or must instead be uniquely designed for subgroups or, at the extreme, for each individual. Foa (1971) has discussed various motivators as being either more universal (e.g., money, goods, information) or more particularistic (e.g., personal attention, social recognition, services). Money and personal attention are at extreme but opposite ends of the particularistic dimension. Foa suggests that money is least particularistic of all motivators because it retains its same value without regard to the relationship between the intervener and the recipient. In contrast, it clearly does matter from whom we receive personal attention for, as Foa points out, its effectiveness is closely linked to the provider. A more particularistic technique would be less generalizable because it would be more context specific.

Another set of context-based issues deal with preexisting conditions. Two moderators have emerged as significant. The first is depth of concern. This concept has proved useful in understanding attitude-behavior relationships. Attitudes are found to be more predictive of behavior when they are held with greater conviction. For example, Abelson (1988) suggests that it is vital to distinguish between those attitudes that people do not genuinely concern themselves about and those that are personally significant for them. There is evidence that success-ful promotion efforts require that people think of an ERB as important from their own point of view (Dwyer et al., 1993; Geller, 1995a, 1995b: Porter, Leeming, & Dwyer, 1995). Motives will be more effective in those instances in which the behavior the motive seeks to promote goes to the core of a person's needs or con-cerns. In contrast, a motive will be ineffective if the behavior being promoted relates to something of less profound importance and thus, more easily ignored if matters or time press.

It is unlikely that a single motive will prove effective on all these dimensions. A durable motive may not be widespread in its appeal. A reliable motive may not be generalizable. The challenge, then, is to identify a broad collection of motives for practitioners to use. In deciding where to direct our attention, it is worth noting that extrinsic motives, as a general class, seem deficient in a number of the evalua-tion criteria (De Young, 1993). There is hope for better outcomes when dealing with intrinsic motives.

Reconsidering a Much Maligned Motive

Self-interest is traditionally identified as a major *source* of environmental problems (Hardin & Baden, 1977; Mansbridge, 1990). This presumption was cen-tral to much of the early research on ERB. It is, for instance, a fundamental part of human behavioral ecology, which argues that humans are egocentric gain-maximizers, having evolved to consume resources with little or no concern for efficiency, to pass waste and costs on to others, and to form small groups that exclude and neglect the interest of others. Self-interest is modeled as focusing solely on short-term individual or familial gain to the exclusion of long-term soci-etal or environmental benefits (Low & Heinen, 1993).

In sharp contrast, research reported this past decade suggests the possibility that self-interest is a potential *solution* to environmental problems. In findings that further support the notion that ERB is multiply determined, Stern, Dietz, and Kalof (1993) argue that self-interest works in concert with altruism to promote ERB, and Fusco (1991) reports that office recycling programs that begin with legal coercion or social concern often continue by adopting a motive that is best described as eco-nomic self-interest.

Recent work on volunteerism speaks to the long-term effect of attending to one's self-interest. Snyder and colleagues, employing a functional approach, report that people have a wide variety of reasons for volunteering, including valu-ing social issues, concern for community well-being, personal development, and esteem enhancement (Clary & Snyder, 1999; Snyder & Omoto, 1992). What is fascinating is that a person with more self-oriented motives (e.g., esteem enhancement, personal development) tends to remain a volunteer longer. In contrast, a person with more community, social-issue-focused, or value-based reasons tends to volunteer for a shorter period. The authors suggest "that the opportunity to have personal, self-oriented, and perhaps even selfish functions served by volunteering was what kept volunteers actively involved" (Omoto & Snyder, 1995, p. 683). If durability is a concern, then these findings suggest that efforts to promote ERB will benefit from attending to the personal benefits derived from such activities.

Before addressing this issue further, it is necessary to clear up two misunderstandings about self-interest. The first involves distinguishing self-interest from selfishness. Self-interest is often devalued as a useful motive because it is, mistakenly, equated with selfishness (Perloff, 1987). It is easy to confuse the two. However, selfishly consuming resources or creating waste without concern for others is quite different from taking care of yourself and maintaining your ability to function effectively in a challenging and frequently chaotic world. The responsibility for getting your own needs met, for gaining a sense of happiness or meaning from life, for maintaining mental vitality and a positive outlook rests only with yourself. If you do take care of yourself and can maintain a positive outlook, then you will be in a much better position to take care of others who cannot take care of themselves (e.g., people who are sick, children) or to advocate for the environment.

A further misunderstanding is the belief that self-interest is only about attaining personal happiness. The extreme of egoism is to believe that the only thing that matters to us is our own happiness and that, by extension, we can never have concern for another person or thing external to us. In their thoughtful book, *Psychology's Sanction for Selfishness*, Wallach and Wallach (1983) clear up this misunderstanding by noting that our individual happiness can depend on what happens to those things about which we care. They state that "we are satisfied or pleased if we attain what we (really) want; we are made happy if something that we (really) wish for comes to pass" (p. 201). Thus, although happiness is experienced personally, it is derived from attaining an outcome, *any* outcome, we care about. A personal sense of satisfaction can be derived from such things as enhancing the well-being of another person or the sustainability of an ecosystem. Framed in this way, self-interest can be tied to a vast number of concerns, many directly relevant to the promotion of ERB and some working with surprising effectiveness.

The Motive of Intrinsic Satisfaction

Research done on intrinsic satisfaction (De Young, 1985, 1986, 1993, 1996) is consistent with the ideas about self-interest presented by Wallach and Wallach (1983). People have reported that certain patterns of behavior are worth engaging in because of the personal, internal contentment that engaging in these behaviors provides. However, these behaviors often focus on issues outside the immediate domain of the self (e.g., protecting the environment, enhancing community). Thus, no ecocentric value need be presumed to account for ERB nor a sociocentric value for helping the community. The ultimate effect may be environmentally or socially beneficial, but the proximate mechanism is self-interest, here in a form called intrinsic satisfaction.

Some researchers have equated intrinsic satisfaction with altruism. If we start with the more traditional definition of altruism, an unselfish concern for others often involving some level of personal sacrifice, and understand intrinsic satisfaction to focus on actions carried out for immediate, personal, and, some might say, self-interested reasons, then clearly they are quite different motives. If, however, an alternate definition is used, namely, that altruism involves getting pleasure from helping behavior, then these are related motives.

The existence and structure of intrinsic satisfactions has emerged over the last 15 years of research on ERB (De Young, 1996). The intrinsic satisfaction categories discussed below emerged from nine studies done during the past decade, with some data published here for the first time. These studies investigated a variety of environmentally responsible behaviors and populations using a common bank of items on intrinsic satisfaction (see Table 1).

Three intrinsic satisfactions are relevant to the discussion of environmental sustainability: (1) satisfaction derived from striving for behavioral competence, (2) frugal, thoughtful consumption, and (3) participation in maintaining a community. A fourth, pleasure from luxuries, was included initially to check for construct

Table 1. Description of the Studies

Study	Date	N	Population studied	Focus of study	Reference
1	1990	159	Food store consumers	Household source reduction	De Young et al., 1993
2	1991	103	Food store consumers	Household source reduction	De Young et al., 1993
3	1991	1,788	Taiwanese office workers	Office recycling	Lee & De Young, 1994
4	1992	73	National Resources Defense Council members	*Mothers and Others* program	
5	1993	169	Environmental Protection Agency employees	Source reduction	Duncan, 1997
6	1995	113	College students[a]	Environmentally responsible behavior	
7	1996	109	College students[b]	Environmentally responsible behavior	
8	1999	396	Homeowners	Reduced consumption and well-being	
9	1999	1,413	Norwegian homeowners	Environmentally responsible behavior	

[a]Random sample of graduate and undergraduate students.
[b]Students from a business school and a school of natural resources.

validity but has produced an interesting finding of its own. In each study the bank of items measuring these intrinsic satisfactions was introduced with a stem question similar to "Please indicate how much satisfaction or enjoyment you get from each of the following items." Participants responded using a 5-point Likert rating scale ranging from *none* to *a very great deal*. Participants rated how much satisfaction they receive from engaging in the activities listed. Factor analysis was used to identify the categories. The items making up these four intrinsic satisfaction categories are reported in Table 2. One fascinating finding to come out of these studies is the coherent, multidimensional nature of intrinsic satisfactions. The participants in the various studies report deriving not a single, all-inclusive sense of satisfaction but numerous and specific satisfactions.

Competence

The first category includes satisfaction derived from striving for behavioral competence. It includes participants' enjoying being able to solve problems and complete tasks. Competence was proposed by White (1959) as a basic human concern, an inclination to strive for ever more effective interactions with the environment. Geller (1995a, 1995b) links competence with ERB when he includes self-efficacy as a major component in his actively caring hypothesis. In White's conceptualization, competence has both a skill and motivational aspect. The studies mentioned here measure not the ability to interact effectively (e.g., assessment of specific skills or expertise) but the motive for developing and maintaining these competencies.

That humans would be motivated to develop behavioral competence is not, on first glance, an impressive finding. What is fascinating, however, is that the participants report deriving personal enjoyment from such effort and that this category has generally been the most highly endorsed of all intrinsic satisfactions.

Frugality

With survival having always depended on the careful stewardship of finite resources, we might expect people to have come to recognize the sorts of lifestyles in which such care was both possible and supported. However, it is not only important for people to recognize such patterns; they should also find them satisfying to pursue. Thus, we could argue that satisfaction from frugality is at the core of ERB.

Once a commonplace virtue (Nash, 1998), frugality is needed now more than ever. Yet, it need not be adopted solely on utilitarian grounds. As measured in these studies, frugality is perceived by the participants as a satisfying activity worth pursuing in it own right. Here we have an excellent instance of what Wallach and Wallach (1983) are arguing for. The positive environmental benefits that pursuing frugal behavior creates for both society and the ecosystem are the direct result of a

Table 2. Intrinsic Satisfaction Categories

Category name and items included		Study 1 1990 N = 159	Study 2 1991 N = 103	Study 3 1991 N = 1,788	Study 4 1992 N = 73	Study 5 1993 N = 169	Study 6 1995 N = 113	Study 7 1996 N = 109	Study 8 1999 N = 396	Study 9 1999 N = 1,413
COMPETENCE	Mean	4.54	4.43	4.17	4.56	4.53	4.07	4.41	4.27	3.87
	Standard deviation	.51	.58	.64	.50	.47	.61	.56	.53	.79
	Alpha	.85	.85	.64	.70	.81	.54	.78	.66	.82
Knowing how to finish a task		●	●	○	●	●				
Remaining competent at meeting life's challenges		●	●	●	●	●	●	●	●	●
Being good at the things I need to do		●	●	●	●	●		●	●	●
Learning how to solve most problems I face		●	●	●	●	●			●	●
Knowing what things I'm good at doing		●	○	●	●	●				
Discovering new things I'm good at doing		○	○	○	○	●				
Knowing the things I'm not competent at doing		○	●							
Possessing many new things										
Having better tools for life's tasks						●				
People would respect me										
FRUGALITY	Mean	4.20	4.12	4.11	4.24	4.05	3.67	3.86	3.47	3.17
	Standard deviation	.67	.73	.53	.72	.56	.77	.60	.73	.74
	Alpha	.80	.89	.82	.79	.71	.75	.73	.77	.78
Finding ways to avoid waste		●	●	●	●	●	●	●	●	●
Keeping something running past its normal life		●	●	●	●	●	●	●	●	●
Finding ways to use things over and over		●	●	●	●	●		●	●	●
Repairing rather than throwing things away				●	●	●				
Saving things I might need someday					○	●				
Consuming a minimum amount of resources						●				
Using technology to do things more efficiently						●				
Developing ways to use resources more effectively						●				
Taking actions that make life more simple						●				

The things I buy would be well suited to the task

Buying items I need from a secondhand shop

Using the library rather than buying new books/mags

		1	2	3	4	5	6	7	8
The things I buy would be well suited to the task								•	
Buying items I need from a secondhand shop								•	•
Using the library rather than buying new books/mags									•
PARTICIPATION	Mean	4.04	4.09	4.21	4.32		3.70		
	Standard deviation	.75	.77	.60	.64		.86		
	Alpha	.80	.84	.81	.85		.76		
Taking actions which can change the world		•	•	•	•		•		
Doing things that help bring order to the world		•	•	○	•		•		
Helping to make sense out of the world		•	•	◐	•		•		
Doing things that matter in the long run		○	○	○	•				
Fitting into our place in the natural scheme of things			○	•					
Influencing how society solves problems					•				
LUXURY	Mean				3.24	3.46	3.03	2.18	2.21
	Standard deviation				.79	.75	.82	.64	.67
	Alpha				.70	.69	.65	.78	.79
Having clothing that is in style					•	•	•	•	•
Having many items to choose from when purchasing					•	•	•	•	•
Having the luxuries and conveniences of our society					•	•	•	•	•
Being a citizen of a country with vast resources					○	•			
Having new items to try, evaluate and buy					•			•	•
Using the latest consumer or electronic gadget								•	•
Being the first to own an unusual product								•	•

Note. Solid circles indicate items that loaded in the factor analysis; open circles indicate items included on the survey instrument but not meeting inclusion criteria. Blanks indicate items not included on the survey instrument.

self-interested focus on achieving personal happiness. An ecocentric orientation is not only consistent with self-interest, it may be derived from it.

Participation

Ellis and Gaskell (1978, as reported in Stern & Gardner, 1981) note that a motive to conserve can come from as subtle a factor as direct participation. In our studies, the participants consistently report deriving satisfaction from participation in community activities and value opportunities to take action that makes a difference in the end.

There is undoubtedly a prosocial inclination in people. This inclination seems quite broad and genuine, not at all calculated. It certainly includes caring about the welfare of other humans and helping them through hard times, but this inclination should not be mistaken for altruism, for it also includes a broader range of concerns (S. Kaplan, this issue). Included is an eagerness to share news, finding pleasure from working with others toward a common goal, and, given the right conditions, a willingness to expend considerable effort in developing positive relations with others and in sharing skills and knowledge. The inclination is as much about interacting with other people as it is about helping them. A central theme here is being needed, of having the chance to make a contribution that is not optional but necessary. It seems that when people discern a role for themselves and become convinced that their efforts truly matter, a powerful motive force is unleashed (S. Kaplan, 1990).

Luxury

The final category focuses on the satisfaction gained from having both the conveniences of our modern society and access to new and novel products. This category captures the satisfaction people derive from being part of a thriving society. Since this category tapped into behaviors that were the opposite of conservation, it was initially included as a means of testing for construct validity. A more useful finding emerged, however.

Early work on ERB suggested that the lifestyle we would soon need to adopt to ensure sustainability would be austere, perhaps even somber. Environmental responsibility was often portrayed as the behavioral equivalent of freezing in the dark. We were told to expect neither comfort nor amenity in a sustainable society. It is in this sense that satisfaction gained from luxuries might be considered to be in conflict with other environmentally compatible satisfactions. However, the participants did not view satisfaction derived from luxury as the antithesis of satisfaction gained from the other behavioral patterns. Although logic might suggest a negative correlation between luxury and the other intrinsic satisfaction categories, no such data have emerged. Thus, there is no inherent conflict between ERB and enjoying a modest level of material well-being (De Young & Kaplan, 1985–86). This is a very hopeful finding, for it suggests that there need not be extensive internal dissonance as people begin a transition from a material-focused to a conservation-focused lifestyle.

Expanding on the Urge Toward Competence

Researchers have explored in detail whether attitude and subjective norms are necessary and sufficient to cause behavior change (Ajzen & Fishbein, 1980). The findings suggest that although attitudes and norms sometimes cause behavior change, their influence is significantly reduced when we consider the effects of other variables, including past experience with the behavior (Ajzen, 1991), increased familiarity with the situation, and skill in carrying out the behavior (Gray, 1985). Without considering these variables, we make the error of assuming that once people know *what* they should do and *why* they should do it, they will automatically know *how* to proceed. The issue here is an essential, underlying, and yet sometimes overlooked aspect of behavior change: the need people have for, and the satisfaction people derive from, a sense of competence.

When White (1959) proposed competence or "effectance" as a fundamental human concern, he was arguing for an evolutionarily derived metamotive. Leff, Gordon, and Ferguson (1974) support this claim and show that the research of De Charms (1968, 1971) and J. W. Brehm (1966; S. Brehm & J. W. Brehm, 1981), as well as reinterpretation of White's own earlier research, provides a strong case for believing that the human concern for competence is a primary source of motivation. White also made claims about the intrinsic nature of competence. He argued that the urge toward competence is self-initiating and self-rewarding (White, 1971) and that behaviors associated with competence are highly focused activities that are, in their essence, intrinsically reinforcing (Wandersman, 1979):

> When this particular sort of activity is aroused in the nervous system, [competence] motiva-tion is being aroused, for it is characteristic of this particular sort of activity that it is selec-tive, directed, and persistent, and that instrumental acts will be learned for the sole reward of engaging in it. (White, 1959, p. 323)

Thus, it is possible that a program built upon competence will achieve the durabil-ity common to intrinsically motivated behavior.

However, when considering the role of competence in behavior change, par-ticular attention should be paid to contextual issues. People find unpleasant and thus avoid situations in which they cannot advance or utilize their competence. When people are not sure how to proceed with a new behavior, they are easily over-whelmed. What seems to others a simple action may become for them a major chal-lenge. The issue here goes well beyond a lack of procedural knowledge. It can involve not even knowing what the right questions to ask are. The study of human behavior documents the negative impact of such a state of affairs (S. Kaplan & R.

Kaplan, 1982); when in such a circumstance people will avoid attempting a new behavior regardless of genuine concern, positive attitude, strong social norm, or external inducement. Yet, it is a mistake to describe such people as unmotivated. They are strongly motivated by a desire to be competent. Unfortunately, in such a circumstance, the most reasonable action for people to take might be to avoid trying anything. By ignoring the role competence plays in behavior change, we may inadvertently create situations that cause not adoption of a new behavior but withdrawal and feelings of helplessness.

On a more positive note, the human urge toward competence may readily explain the conditions under which people will consider adopting ERB. It may be no more complicated than providing a context in which procedural information is readily available and behavior can tentatively be tried in a supportive environment. Such a situation would allow people to fulfill an innate desire to utilize and enhance their competence.

Conclusion

If, as White argues, competence is a fundamental motive, then it should sometimes be apparent in the content of other motives. For instance, we might ask whether it is possible to reframe the intrinsic satisfaction categories of frugality and participation as issues of competence. In fact, both do contain the notion of developing skills and abilities useful in taking care of the planet, at either the global or the local scale. Frugality involves resource competence. Being proficient at making things last is reported by the study participants as a valued skill. Participation contains the theme of being effective at making a difference in one's community. There is satisfaction gained from being capable of bringing order to chaos. Perhaps we might build upon the intrinsic satisfaction people gain from being competent at doing things that have a positive effect in a larger context and that matter in the long run.

Similarly, we should expect to find evidence that the urge toward competence is predictive of ERB. Such evidence is emerging. In a study of observed ERB, procedural knowledge was effective at differentiating known conservers from nonconservers (De Young, 1988–89), and a study of household reuse and recycling behavior found that although beliefs predicted self-reported ERB, competence successfully predicted observed ERB (Corral-Verdugo, 1997).

The next challenge will be discovering how to use intrinsic satisfactions to promote ERB. Reichel and Geller (1981) began this quest when they suggested that if we expect and value ERB, then "such norms may even be internalized by individuals so that conserving behaviors become intrinsically reinforced" (p. 88). There is evidence that people's intrinsic motives to conserve can be nurtured and developed. Vining and Ebreo (1990) report that ERB can shift from being initiated and maintained by extrinsic motives toward being influenced by intrinsic motives.

In a fascinating study that speaks directly to the durability issue, Werner and Makela (1998) report that those individuals who actively reframed ERB to emphasize the derived satisfaction were more likely to conserve on both a short- and long-term basis.

In conclusion, it is clear that no single motive is optimal for promoting ERB. No motive has universal appeal, works under all conditions or in all situations. No motive is likely to meet both short- and long-term goals. The widespread promotion of ERB will require an understanding of the great diversity of motives people find acceptable and empowering. Yet, given that there are a huge number of environmentally responsible behaviors that will need to be encouraged, it seems prudent to explore those techniques that score well on the durability and generalizability dimensions.

References

Abelson, R. P. (1988). Conviction. *American Psychologist, 43*, 267–275.

Ajzen, I. (1991). The theory of planned behavior. *Organizational Behavior and Human Decision Processes, 50*, 179–211.

Ajzen, I., & Fishbein, M. (1980). *Understanding attitudes and predicting social behavior.* Englewood Cliffs, NJ: Prentice-Hall.

Allen, J. B., & Ferrand, J. L. (1999). Environmental locus of control, sympathy, and proenvironmental behavior: A test of Geller's actively caring hypothesis. *Behavior and Environment, 31*, 338–353.

Brehm, J. W. (1966). *A theory of psychological reactance.* New York: Academic Press.

Brehm, S., & Brehm, J. W. (1981). *Psychological reactance: A theory of freedom and control.* New York: Academic Press.

Brickman, R., & Campbell, D. T. (1971). Hedonic relativism and planning the Good Society. In M. M. Appley (Ed.), *Adaptation-level theory: A symposium* (pp. 287–303). New York: Academic Press.

Clary, E. G., & Snyder, M. (1999). The motivations to volunteer: Theoretical and practical considerations. *Current Directions in Psychological Science, 8*, 156–159.

Cone, J. D., & Hayes, S. C. (1980). *Environmental problems/behavioral solutions.* Monterey, CA: Brooks/Cole.

Cook, S. W., & Berrenberg, J. L. (1981). Approaches to encouraging conservation behavior: A review and conceptual framework. *Journal of Social Issues, 37*, 73–107.

Corral-Verdugo, V. (1997). Dual "realities" of conservation behavior: Self-report vs. observations of re-use and recycling behavior. *Journal of Environmental Psychology, 17*, 135–145.

De Charms, R. (1968). *Personal causation: The internal affective determinants of behavior.* New York: Academic Press.

De Charms, R. (1971). From pawns to origins: Toward self-motivation. In G. S. Lesser (Ed.), *Psychology and educational practice* (pp. 380–407). Glenview, IL: Scott, Foresman.

De Young, R. (1985). Encouraging environmentally appropriate behavior: The role of intrinsic motivation. *Journal of Environmental Systems, 15*, 281–292.

De Young, R. (1986). Some psychological aspects of recycling: The structure of conservation satisfactions. *Environment and Behavior, 18*, 435–449.

De Young, R. (1988–89). Exploring the difference between recyclers and non-recyclers: The role of information. *Journal of Environmental Systems, 18*, 341–351.

De Young, R. (1993). Changing behavior and making it stick: The conceptualization and management of conservation behavior. *Environment and Behavior, 25*, 485–505.

De Young, R. (1996). Some psychological aspects of a reduced consumption lifestyle: The role of intrinsic satisfaction and competence. *Environment and Behavior, 28*, 358–409.

De Young, R., Duncan, A., Frank, J., Gill, N., Rothman, S., Shenot, J., Shotkin, A., & Zweizig, M. (1983). Promoting source reduction behavior: The role of motivational information. *Environmental and Behavior, 25*, 70–85.

De Young, R., & Kaplan, S. (1985–86). Conservation behavior and the structure of satisfactions. *Journal of Environmental Systems, 15*, 233–242.

Duncan, A. P. (1997). *Source reduction in context: A conceptual framework and field study of waste prevention behavior.* Unpublished doctoral dissertation, University of Michigan, Ann Arbor.

Dwyer, W. O., Leeming, F. C., Cobern, M. K., Porter, B. E., & Jackson, J. M. (1993). Critical review of behavioral interventions to preserve the environment: Research since 1980. *Environment and Behavior, 25*, 275–321.

Eckersley, R. (1992). *Environmentalism and political theory: Toward an ecocentric approach.* Albany, NY: State University of New York Press.

Foa, U. G. (1971). Interpersonal and economic resources. *Science, 171*, 345–351.

Fusco, M. A. C. (1991). Recycling in the office initially may be motivated by altruism, but ultimately such efforts continue because they are cost-effective. *Employment Relations Today, 17*, 333–335.

Geller, E. S. (1987). Applied behavior analysis and environmental psychology: From strange bedfellows to a productive marriage. In D. Stokols & I. Altman (Eds.), *Handbook of environmental psychology* (pp. 361–388). New York: Wiley.

Geller, E. S. (1989). Applied behavior analysis and social marketing: An integration for environmental preservation. *Journal of Social Issues, 45*, 17–36.

Geller, E. S. (1992). Solving environmental problems: A behavior change perspective. In S. Staub & P. Green (Eds.), *Psychology and social responsibility: Facing global challenges* (pp. 248–268). New York: New York University Press.

Geller, E. S. (1995a). Actively caring for the environment. *Environment and Behavior, 27*, 184–195.

Geller, E. S. (1995b). Integrating behaviorism and humanism for environmental protection. *Journal of Social Issues, 51*, 179–195.

Geller, E. S., Winett, R. A., & Everett, P. B. (1982). *Preserving the environment: New strategies for behavioral change.* New York: Pergamon Press.

Gray, D. B. (1985). *Ecological beliefs and behaviors: Assessment and change.* Westport, CT: Greenwood Press.

Hardin, G., & Baden, J. (1977). *Managing the commons.* San Francisco: W. H. Freeman.

Kaplan, A. W. (1999). From passive to active about solar electricity: Innovation decision process and photovoltaic interest generation. *Technovation, 19*, 467–481.

Kaplan, S. (1990). Being needed, adaptive muddling and human-environment relationships. In R. I. Selby, K. H. Anthony, J. Choi, & B. Orland (Eds.), *EDRA 21.* Oklahoma City, OK: Environmental Design Research Association.

Kaplan, S., & Kaplan, R. (1982). *Cognition and environment: Functioning in an uncertain world.* New York: Praeger.

Katzev, R. D., & Johnson, T. R. (1987). *Promoting energy conservation: An analysis of behavioral research.* Boulder, CO: Westview Press.

Kreutzwiser, R. (1991). Municipal recycling programs and household conservation behaviour (Project ER 508G). Department of Geography, University of Guelph, Guelph, Ontario, Canada.

Lee, Y., & De Young, R. (1994). Intrinsic satisfaction derived from office recycling behavior: A case study in Taiwan. *Social Indicators Research, 31*, 63–76.

Lee, Y., De Young, R., & Marans, R. W. (1995). Factors influencing individual recycling behavior in office settings: A study of office workers in Taiwan. *Environment and Behavior, 27*, 380–403.

Leff, H. L., Gordon, L. R., & Ferguson, J. G. (1974). Cognitive set and environmental awareness. *Environment and Behavior, 6*, 395–447.

Lepper, M. R. (1981). Intrinsic and extrinsic motivation in children: Detrimental effects of superfluous social controls. In W. Collins (Ed.), *Aspects of the development of competence: The Minnesota Symposium on Child Psychology* (Vol. 14, pp. 155–160). Hillsdale, NJ: Erlbaum.

Lepper, M. R., & Greene, D. (Eds.). (1978). *The hidden costs of rewards: New perspectives on the psychology of human motivation.* Hillsdale, NJ: Lawrence Erlbaum.

Low, B. S., & Heinen, J. T. (1993). Population, resources, and environment: Implications of human behavioral ecology for conservation. *Population and Environment, 15*, 7–41.

Mansbridge, J. J. (1990). *Beyond self interest*. Chicago: University of Chicago Press.

Mazis, M. R. (1975). Antipollution measures and psychological reactance theory: A field experiment. *Journal of Personality and Social Psychology, 31*, 654–660.

McClelland, L., & Canter, R. J. (1981). Psychological research on energy conservation: Context, approaches, methods. In A. Baum & J. E. Singer (Eds.), *Advances in environmental psychology: Vol. 3. Energy conservation: Psychological perspectives* (pp. 1–25). Hillsdale, NJ: Lawrence Erlbaum.

Midgley, M. (1978). *Beast and man: The roots of human nature*. New York: New American Library.

Nash, J. A. (1998). On the subversive virtue: Frugality. In D. A. Crocker & T. Linden (Eds.), *Ethics of consumption: The good life, justice and global stewardship* (pp. 416–436). New York: Rowman & Littlefield.

Omoto, A. M., & Snyder, M. (1995). Sustained helping without obligation: Motivation, longevity of service, and perceived attitude change among AIDS volunteers. *Journal of Personality and Social Psychology, 68*, 671–686.

O'Neal, E. C., Kipnis, D., & Craig, K. M. (1994). Effects on the persuader of employing a coercive influence technique. *Basic and Applied Social Psychology, 15*, 225–238.

Oskamp, S., Harrington, M. J., Edwards, T. C., Sherwood, T. L., Okuda, S. M., & Swanson, D. C. (1991). Factors influencing household recycling behavior. *Environment and Behavior, 23*, 494–519.

Perloff, R. (1987). Self-interest and personal responsibility redux. *American Psychologist, 42*, 3–11.

Porter, B. E., Leeming, F. C., & Dwyer, W. O. (1995). Solid waste recovery: A review of behavioral programs to increase recycling. *Environment and Behavior, 27*, 122–152.

Reich, J. W., & Robertson, J. L. (1979). Reactance and norm appeal in antilittering messages. *Journal of Applied Social Psychology, 9*, 91–101.

Reichel, D. A., & Geller, E. S. (1981). Applications of behavioral analysis for conserving transportation energy. In A. Baum & J. E. Singer (Eds.), *Advances in environmental psychology: Vol. 3. Energy conservation: Psychological perspectives* (pp. 53–91). Hillsdale, NJ: Lawrence Erlbaum.

Rind, B., & Kipnis, D. (1999). Changes in self-perceptions as a result of successfully persuading others. *Journal of Social Issues, 55*, 141–156.

Schwartz, S. H. (1970). Moral decision making and behavior. In J. Macauley & L. Berkowitz (Eds.), *Altruism and helping behavior* (pp. 127–141). New York: Academic Press.

Schwartz, S. H., & Howard, J. A. (1981). A normative decision-making model of altruism. In J. P. Rushton & R. M. Sorrentino (Eds.), *Altruism and helping behavior: Social, personality, and developmental perspectives*. Hillsdale, NJ: Lawrence Erlbaum.

Sieber, S. D. (1981). *Fatal remedies: The ironies of social intervention*. New York: Plenum Press.

Snyder, M., & Omoto, A. M. (1992). Who helps and why? The psychology of AIDS volunteerism. In S. Spacapan & S. Oskamp (Eds.), *Helping and being helped: Naturalistic studies* (pp. 213–239). Newbury Park, CA: Sage.

Stern, P. C., & Dietz, T. (1994). The value basis of environmental concern. *Journal of Social Issues, 50*, 65–84.

Stern, P. C., Dietz, T., & Kalof, L. (1993). Value orientations, gender, and environmental concern. *Environment and Behavior, 25*, 322–348.

Stern, P. C., Dietz, T., Kalof, L., & Guagnano, G. (1995). Value, beliefs, and environmental action: Attitude formation toward emergent attitude objects. *Journal of Applied Social Psychology, 25*, 1611–1636.

Stern, P. C., & Gardner, G. T. (1981). Psychological research and energy policy. *American Psychologist, 36*, 329–342.

Thompson, S. C. G., & Barton, M. A. (1994). Ecocentric and anthropocentric attitudes toward the environment. *Journal of Environmental Psychology, 14*, 149–157.

Vining, J., & Ebreo, A. (1990). What makes a recycler? A comparison of recyclers and non-recyclers. *Environment and Behavior, 22*, 55–73.

Wallach, M. A., & Wallach, L. (1983). *Psychology's sanction for selfishness: The error of egoism in theory and therapy*. San Francisco: W. H. Freeman.

Wandersman, A. (1979). User participation: A study of types of participation, effects, mediators, and individual differences. *Environment and Behavior, 11*, 185–208.

Werner, C. M., & Makela, E. (1998). Motivations and behaviors that support recycling. *Journal of Environmental Psychology, 18*, 373–386.

White, R. W. (1959). Motivation reconsidered: The concept of competence. *Psychological Review, 66*, 297–333.

White, R. W. (1971). The urge towards competence. *American Journal of Occupational Therapy, 25*, 271–274.

RAYMOND DE YOUNG is an Associate Professor of Environmental Psychology and Conservation Behavior in the University of Michigan School of Natural Resources and Environment. His research focuses on the psychology of environmental stewardship, particularly the role of intrinsic motivation in promoting conservation behavior. His current work is exploring the effect that mental (attentional) vitality has on the promotion of psychological well-being and environmentally responsible behavior and includes the study of the restorative effects of time spent in natural settings.

Journal of Social Issues, Vol. 56, No. 3, 2000, pp. 527–541

The Application of Persuasion Theory to the Development Of Effective Proenvironmental Public Service Announcements

Renee J. Bator*

State University of New York, Plattsburgh

Robert B. Cialdini

Arizona State University

The goal of this article is to provide specific guidelines to help create effective proenvironmental public service announcements (PSAs). Campaign designers are encouraged to initially identify and investigate the optimal target audience and then draft and test reactions by samples of that audience using pilot messages. Designers are also advised to consider research on attitude persistence, memory, and social norms and apply this research to the message content and presentation style. The article concludes with an application of research from social psychology to a series of overall guidelines for effective PSAs. If environmental campaign developers follow these specifications, the chance of PSA success should be enhanced.

Public service announcements (PSAs) are designed to inform or induce certain behaviors in specific audiences, generally for noncommercial profit using mass media-approaches (adapted from Rogers & Storey, 1987, by Rice & Atkin, 1989). The advantage of using PSAs to promote prosocial behavior is due in part to their ability to efficiently and repeatedly penetrate a large target population, with the possibility of relying on highly respected sources as spokespersons (Hornik, 1989). One of the best-known environmental PSAs was presented in the 1970s.

*Correspondence concerning this article should be addressed to Renee Bator, Psychology Department, 101 Broad Street, SUNY Plattsburgh, Plattsburgh, NY 12901 [e-mail: renee.bator@plattsburgh.edu].

The Iron Eyes Cody messages featured a heavily littered environment and a Native American with a tear in his eye and the slogan, "People start pollution; people can stop it." Since its original airing in 1971, the PSA has been seen by an estimated 50 million Americans (Dwyer, 1999). In 1997, it was named one of the top 50 commercials of all time by *Entertainment Weekly* (Dwyer, 1999). Although this message certainly elicits emotional reactions from viewers, this article will point out a possible flaw in the message, along with suggestions for creating an even stronger PSA.

Although there is a great deal of persuasion research that addresses attitude change and corresponding behavior change, PSAs are typically designed without taking advantage of this information. Proenvironmental campaigns face a special problem, because the messages attempt to direct a behavior that does not occur until a later time. The goal of this article is to provide specific guidelines to help create effective proenvironmental PSAs.

When developers of proenvironmental PSAs neglect to consider basic principles derived from mass-media communications research, their efforts to bring about behavioral change are likely to be unsuccessful. A great deal of research has examined the importance of identifying a target audience, learning about their attitudes and behaviors related to the target issue, and then pilot testing responses to preliminary versions of the message. Based on previous research, Mendelsohn (1973) found that public information campaigns have a relatively high probability of success if (1) campaign developers assume that most audiences are likely to be only mildly interested in the message, (2) middle-range goals are set (e.g., developers feel confident that simple message exposure will lead to the desired information gain or change in behavior), and (3) the target audience is thoroughly investigated in terms of demographics, lifestyles, values, and mass-media habits. Mendelsohn described three information campaigns that were highly successful because each of their designs reflected close collaboration between social scientists and communications specialists.

Atkin and Freimuth (1989) provide a step-by-step guide to formative evaluation research in campaign designs. They contend that evaluation research should first answer questions about audience attitudes and behaviors prior to the campaign design, then evaluate the design's execution and effectiveness during and after a campaign (Flay & Best, 1982; Flay & Cook, 1989). This evaluation research process includes two major steps, the preproduction stage and the pretesting stage, each of which has numerous substeps. In the preproduction stage of the research, the strategist attempts to discover as much as possible about the target audience before specifying objectives, drafting strategies, and matching the message to the audience. Then the pretesting stage involves the process of methodically collecting reactions from the intended audience based on preliminary versions of messages before they are finalized (Bertrand, 1978; U.S. Department of Health and Human Services, 1984). Since the current process of producing PSAs (based mainly on creative inspiration) has achieved only limited success, this proposal for step-by-step research certainly merits consideration by managerial and creative personnel.

McGuire (1989) also provided useful guidelines for creating effective public communication campaigns. He described how fundamental theories about a person's structure and motivation affect that person's response to a persuasive message. An input-output matrix was formulated to better understand the communication variables (input) and the response steps (output). Along the input axis are important aspects of the message such as the source (age, gender, socioeconomic status, ethnicity, credibility, and attractiveness), message factors (delivery style, length, repetition, speed of speech, and vividness), channel factors (television, radio, newspaper, or magazine, as well as specifics within these), receiver factors (age, education, gender, lifestyle), and finally target behaviors at which the communication is aimed—these are called the destination factors (immediate versus long-term change, trying to encourage a new behavior or stop a current habit). Most of these input options are under the control of the campaign developers, and thus they can be manipulated to achieve the most effective response.

The output factors include 12 consecutive response substeps that are all considered necessary if the communication is to be practical. For example, the public must have contact with the message and, having been exposed to it, must pay attention to it, like it, understand and learn from its content, agree with it, store the information and be able to retrieve it later, and make decisions based on it. The final steps include acting from that decision, getting reinforced for such actions, and taking on postcompliance activities (such as convincing others or restructuring one's self-concept) that fortify the new attitude inspired by the message.

McGuire's matrix enables producers of proenvironmental PSAs to consider design alternatives such as whether the spokesperson should be male or female and whether to use a soundtrack or just show the source's face speaking, as a function of how each option may be responded to on the output matrix. Although Atkin and Freimuth (1989) also focused on the importance of design evaluation, they emphasized only a few responses such as liking or recall. McGuire realized that it is important to complete a thorough evaluation of potential PSAs that goes beyond simply picking the one that is better liked. The current "Got Milk?" advertising campaign is under scrutiny because the highly popular and expensive celebrity advertisements promoting milk consumption have correlated with decreased sales (Leonhardt, 1998). Although this is only correlational data, it does indicate that liking a message does not necessarily lead to the desired behavior change. McGuire warned that making a decision to use a message with only limited feedback from a target audience may mean choosing a PSA that will not be responded to on the final steps, such as behaving in accord with the message.

McGuire's 12 stages provide a practical checklist for creating and evaluating the effectiveness of proenvironmental PSAs. This checklist could be incorporated along with Atkin and Freimuth's (1989) step-by-step plan to produce a comprehensive evaluation of potential proenvironmental communications. For instance, researchers working on antilittering PSAs could use surveys and focus groups along the lines of Atkin and Freimuth's preproduction stage to measure the effectiveness of McGuire's "receiver" factors and determine potential "message" types and "channel" possibilities. Then campaign designers could turn to the pretesting phase and collect reactions to preliminary antilittering messages using McGuire's output axis for the dependent measures. Those communications that are favorably responded to at the later steps (e.g., antilittering behavior) should be most effective when the message is presented publicly. This process of closely examining the audience and their reaction to preliminary messages is consistent with the advice of McKenzie-Mohr in this issue. He found that social marketing strategies were much more effective if the developers thoroughly considered the target audience and any potential barriers they might have to accepting the message. Once these barriers are recognized, campaign developers can take a more informed approach to designing a successful campaign.

Attitude Persistence Induced by Systematic Processing

Researchers have consistently found that once an individual is exposed to a message, it is how the individual processes the information that determines if persuasion will be enduring. Cook and Flay (1978) found that participants who thoughtfully considered message content demonstrated more enduring attitude change; in contrast, when participants had little motivation and/or ability to think about the message presented, the effects were typically short lived. Petty and Cacioppo (1981) developed the elaboration likelihood model of persuasion and argued that the course of persuasion is based on how much mental processing or elaboration the target undergoes. This elaboration likelihood model includes both a peripheral route to persuasion and a central route to persuasion, depending on the target's motivation and ability to process the message.

The peripheral route is used whenever the target's motivation or ability to think about an issue is low. In this case, persuasive influences are more tangential to the issue at hand. The message recipient tends to focus on source characteristics or potential rewards for abiding by the message rather than on the message content. For example, an individual who has not considered using public transportation may notice only the appearance of the actors or the catchy jingle in a public transportation PSA. Any resulting persuasion would not be due to actively considering the issue but instead would be a result of peripheral "persuasion cues" (e.g., source attractiveness). The use of peripheral persuasion cues is rarely successful in producing change when the target audience has prior knowledge of or interest in the issue. This means that a target audience that has been considering the use of public transportation is unlikely to be persuaded in this direction simply by the use of an attractive spokesperson. Moreover, attitude change that does come about through the peripheral route is rarely permanent.

The peripheral route is risky for any type of public communication campaign. The first problem is that most proenvironmental behaviors advocated by PSAs are likely to be thought-provoking, and the peripheral route depends on heuristics. It is more successful at getting people to choose brand A over brand B than at getting them to change their lifestyle. Although peripherally processed behavior change is usually short-term, there are means by which it could produce a more enduring result. For instance, if the target audience is motivated to take public transportation because they see a popular spokesperson advocating this behavior, the targets might begin to consider the advantages of public transportation, and in such a case the final result of peripheral processing could be long-term behavior change. This possibility is consistent with Bem's (1972) self-perception theory, in which people observe their behavior and then infer internal reasons for it.

Central processing usually occurs when the audience for a persuasive message is motivated and able to take the time to consider its content. Central processing is more likely when the issue at hand is personally relevant to the audience. For the central route to be successful, the arguments must be attended to, understood, and integrated into established belief structures. If the communicator is concerned about the target's ability to pay attention to or understand the message, certain tactics can be used to increase the probability that the target will do both of these things, such as repeating the message or providing a written version. When the message's arguments are strong, this active processing can generate positive evaluations and result in the most enduring attitude change. This is said to occur via the following sequence (Petty & Cacioppo, 1981):

Attention → Comprehension → Elaboration → Integration → Enduring Attitude Change

To create a proenvironmental PSA that leads to enduring attitude/behavior change, the central route is probably the best option, as it strives to meet each of the output points on McGuire's matrix. This is certainly easier said than done.

One of the first factors a communication designer should consider is the background attitudes and behaviors of the target audience. This can be done with surveys or focus groups, in accord with Atkin and Freimuth's (1989) preproduction stage. Researchers should gain insight into how meaningful the issue is to the target audience. If the issue is gauged to be important, then central processing is likely to occur. In this case the message should be effective to the degree that the audience favorably responds to it along McGuire's output axis. Of course, simply attending to a PSA is not enough; the target audience must also remember it later, and thus the issue of human memory will now be explored in regard to PSA designs.

Memory

Like advertisements and other types of PSAs, proenvironmental PSAs are not presented at the specific point in time when the target should respond to the message; instead, they are presented with the idea that the target will respond later with an appropriate behavior change. To determine guidelines that will enhance the effects of these PSAs, it is important to understand how the human memory system works. There are a number of topics within the study of memory that have an impact on these specifications. Message designers must be aware that PSAs will be presented in a message-dense environment, so they should consider methods to overcome distraction. The next section of this article investigates how humans store information in memory. If campaign developers are aware of how people organize information and how vivid information stands out in memory, then proenvironmental PSAs can be designed and presented so their message content is easily accessible. From here retrieval becomes an important issue. Including an encoding cue in a message is one way to increase the likelihood that the message content will be recalled at the appropriate time and place in the future. Because PSAs are presented amid numerous other communications, the first feature of memory to be discussed is distraction.

Distraction

Petty and Cacioppo's elaboration-likelihood model predicts that personally involving messages increase thoughtful consideration of the message and strengthen resulting attitude change. In fact Petty, Cacioppo, Haugtvedt, and Heesacker (1986; reported in Petty & Cacioppo, 1986) found that involving messages resulted in more-durable attitude change. But what happens to involvement and persistence when the target's ability to dedicate full attention to the message is taxed?

Pratkanis and Greenwald (1993) investigated how to overcome the obstacle of persuading people in an atmosphere that is cluttered with competing messages. Such an environment hampers message recipients' ability to devote their attention to an issue and results in a great deal of associative interference and forgetting (Baumgardner, Leippe, Ronis, & Greenwald, 1983; Webb & Ray, 1979). Previous research has indicated that consumers in such an environment tend to overlook most product claims and respond only to the key elements in certain personally involving messages (Baumgardner et al., 1983; Chaffee & McLeod, 1973). Pratkanis and Greenwald developed a message utility hypothesis to address this issue. In their consumer marketing approach, participants spent more time considering messages that made positive evaluations of products that were relevant to their purchasing task. It was these useful messages that had the most persistent effects on the participants.

Pratkanis and Greenwald concluded with two suggestions. First, they warned that in a message-dense environment, persuasive appeals that represent uninvolving issues have very limited impact. To counteract this problem the authors recommended using a highly credible source and relying on a single, well-placed, very positive message.

Not surprisingly, many researchers have found highly credible sources to be more persuasive than those with less credibility (Hovland & Weiss, 1951; Kelman & Hovland, 1953). For instance, a PSA that promotes public transportation will probably be more effective if the spokesperson or sponsoring foundation is considered to be either an expert on taking a train or bus, perhaps a regular commuter, or if the target audience finds this individual or organization trustworthy. In this latter case developers probably would want to avoid using employees of the transportation company as spokespersons, since these individuals have a financial interest in promoting buses or trains. Campaign designers should rely on Atkin and Freimuth's (1989) preproduction guidelines to determine which potential sources are considered to be the most credible by the target audience.

Pratkanis and Greenwald's (1993) second suggestion emphasized the elaboration-likelihood model. In an environment swamped with competing communications, it is the involvement of the target that enhances his or her attention to and elaboration of the persuasive appeal. Involvement will be based on whether the message comes across as personally useful to the target. Messages that help resolve specific goals will have the most persistent effects. Pratkanis and Greenwald's research counsels developers of public transportation PSAs that a communication should not just tell members of the public that they should take the train or the bus, but rather it should explain specifically how to use this means of transportation and describe the many ways this behavior will be beneficial.

Storage

Although simply getting the audience's attention is a difficult task, an equally challenging process is providing the message content of a PSA in such a way that audiences will readily store it in memory. Keller (1987) noted that because consumers probably do not make decisions during exposure to an advertisement, consumer memory for advertising is important. As noted previously, proenvironmental PSAs are like advertisements in that they typically encourage a new behavior that will probably not be relevant until a later point in time. Therefore it is important that message creators understand the process of long-term memory storage.

Wood's (1982) research on memory found that retrieval of our past experiences and behaviors is a more important contributor to our current opinions than is retrieval of abstract cognitions. Wood, Kallgren, and Preisler (1985) found that participants with access to background attitudes and behaviors in memory changed their opinion on several target issues to the extent that the messages included high-quality arguments. They considered their results to support the elaboration-likelihood model; high-retrieval participants based their opinion change on the message's content, whereas low-retrieval participants employed more heuristic strategies (Chaiken, 1980). This suggests that campaign designers should survey the target audience to learn about their attitudes and experiences with the persuasive topic. If a common theme becomes apparent, then this "personal" experience should be included in the message. For instance, suppose campaign designers working on a home energy conservation PSA learn that many people in the target audience want to conserve energy but do not know which household appliances and heating and cooling systems use the most energy. If this is a common experience, then a discussion of this issue should be present in the message. The communication should of course continue with a description of simple ways to reduce energy consumption, perhaps by explaining the savings associated with reducing the temperature on the hot water heater or running the dishwasher only when it is full. When the target audience sees this comparison of household appliances in the message, it should make the issue of energy conservation more personally relevant, and the central route to persuasion should be more likely to become activated.

Communicators have relied on concrete, vivid messages to be more personally meaningful, more emotionally arousing, and as a result more influential. Nisbett and Ross (1980) defined vivid information as likely to draw and hold our attention and to stimulate the imagination to the extent that it meets three criteria: (1) it grabs us emotionally, (2) it is specific and triggers our imagination, and (3) it is immediate in a sensory, temporal, or spatial way.

Rhoads (1994) investigated this issue to determine what type of vivid message is most persuasive. He predicted that an effective message should vividly portray its thesis but avoid adding vivid, irrelevant details. Rhoads found that a message that emphasized the main point with vivid details, without vivifying extraneous details, was rated more positively than messages that vivified irrelevant details in terms of liking, interest, and agreement. He speculated that adding extraneous details might undermine the persuasiveness of messages by distracting the participants from the main point of the communication. Based on this research, an effective PSA should demonstrate the main argument of the message with a vivid description while avoiding vivid surrounding details that may distract from the message. For instance, an antilittering PSA that demonstrates social disapproval of littering should not present attention-drawing versions of a park environment (the beach, the swimmers, or the attractive landscape), but rather it should vividly demonstrate how people socially disapprove of a litterer.

Retrieval Cues

Recall of a message is crucial if the target is going to respond to it at the appropriate time in the desired manner. One group of researchers studied more than 200 energy conservation programs of California utility companies and found that the recall rates for these messages were often as low as 7% (Condelli et al., 1984). Considerable research has investigated this issue from an advertising perspective. Keller (1987) noted that because consumer decisions do not usually take place during exposure to an advertisement, memory is an important aspect of successful advertising. Advertising encoding cues are pieces of verbal or visual information that are initially included in an advertisement that establish a connection between the brand and the advertisement's message. Keller's research on memory factors in advertising opens with an industry example of encoding/retrieval cues that he states helped motivate his research. In the 1970s, Quaker Oats had a highly popular advertising campaign for Life cereal with its classic "Mikey" commercial. To profit from this well-liked message, Quaker included a small, still photo from the commercial on the front of the Life cereal package. Researchers have since found that effective recall is enhanced by strong similarity between the stimulus in the message and a matching (retrieval) cue in the natural environment (Craik, 1981; Tulving, 1979). Retrieval cues in the behavioral setting enhance recall of brand claims and lead to more positive brand reactions (Keller, 1987).

Because of the time lag between exposure to a proenvironmental PSA and the desired action of the target audience, retrieval cues can be a great aid. Tulving and Osler (1968) tested recall of word pairs and found that recall increased dramatically for participants whose memory task included a matching encoding cue and retrieval cue compared to those whose cues did not match. Fisher and Craik (1977) found that this recall is even more impressive when the encoding cue requires effortful processing. Bator (1997) applied these findings to a test of the effectiveness of various amateur antilittering PSAs. The least amount of littering occurred among participants who saw a PSA that emphasized social disapproval of littering during a close-up of a piece of litter. These findings suggest that campaign developers should consider what visual cues are likely to be in the natural environment at the time the target audience is going to act on the message. Such a cue should be emphasized in the message by focusing on it while the most persuasive statements are made. When the target sees this cue at the time of a behavioral decision, it should spark recall of the message's main point.

After deciding on an encoding cue, the next question is what main point should be stated during a focus of this cue. Cialdini, Kallgren, and Reno (1991) executed a research program that is highly applicable to this article. They opened their report of this work by describing the antilittering PSA from the 1970s known as the "Iron Eyes Cody spot." In this message a Native American paddles a canoe up a litter polluted river and then tearfully watches as a bag of trash is thrown on the side of a littered highway from a passing car window. Cialdini and his colleagues explained that although the message certainly emphasized that littering is wrong, at the same time it carried a subtext implying that littering is commonplace. Descriptive norms provide information about what most people do, and the Iron Eyes Cody message inadvertently emphasized the descriptive norm that many people litter. Injunctive norms provide information about social approval and disapproval, and the Iron Eyes Cody spot gets most of its persuasive appeal through the strong emphasis on social disapproval of littering. Cialdini and his colleagues tested descriptive versus injunctive norms in field settings to determine which one is a more powerful motivator of antilittering behavior. They found that injunctive social norms were the most widely applicable in their ability to encourage specific behaviors across a variety of situations and target populations.

Based on this norm research, developers of proenvironmental PSAs are reminded to use injunctive norms by showing social approval for the desired behavior and/or social disapproval for failing to follow this desired behavior. PSA developers are also advised to avoid the common practice of showing many people engaging in the behavior that is being discouraged. For instance, a recycling PSA could portray everyone in an office recycling papers and strongly disapproving of the one person who throws paper into the trash. This kind of message could also incorporate a retrieval cue by emphasizing the disapproval during a close-up of a wadded up piece of paper. The message may also want to explain the ease and benefits of recycling but should avoid pointing out any information about problems with recycling or the low number of people who recycle, as these latter points inadvertently stress an antirecycling descriptive norm. A recycling PSA that emphasizes prorecycling descriptive and injunctive norms and includes an encoding cue should boost recall of the message at a later point when the desired behavior becomes relevant and increase the persuasive power of the campaign.

Commitment and Consistency

Researchers have found that when individuals feel committed to a certain behavior, they will often adopt an identity that is consistent with that behavior, the result of which frequently is long-lasting attitude and behavior change. Several theories in social psychology depend on our desire to be and be seen as consistent. Festinger's cognitive dissonance theory (1957) explains how we will change our attitudes in an effort to reduce the discomfort associated with experiencing an inconsistency. Heider's balance theory (1946) describes how we prefer to experience balance among our relationships and will change our attitudes about those in our social circle to maintain balance. As Cialdini (1993) explains, "Once we make a choice or take a stand, we will encounter personal and interpersonal pressures to behave consistently with that commitment" (p. 51).

Pallak, Cook, and Sullivan (1980) examined how commitment affects a person's ability to be persuaded toward a change in energy consumption. They provided a sample of Iowa homeowners with conservation tips, asked them to reduce consumption, and also created commitment in the following way: They informed these people that the names of conservers would be published in the newspaper.

Within a month this experimental group distinguished themselves from control groups (who were not promised publicity or who were not contacted) by saving significantly more gas (an average of 433 cubic feet of natural gas apiece). The experimental group was then told that the publication of names would not be possible and that the experiment had concluded. Although there was never any publicity, the experimental group maintained their energy conservation throughout the following winter months. It seems that the prospect of having their names published in the newspaper was the impetus behind the participants' initial behavior change, but when the publicity did not occur, the new behavior served a committing function, and energy conservation continued.

Katzev and Johnson (1984) have confirmed these results in a similar study. In their experiment, homeowners who signed a written commitment to try to reduce their energy usage conserved significantly more energy than both a control group and homeowners who were offered money for conservation. Commitment is considered to be the catalyst that drives individuals to experience an internal conviction for both a new identity and the corresponding behavior. Once an individual internalizes this conviction, the identity and behavior can continue even beyond the duration of the commitment.

Although it may initially seem that PSAs are poor mechanisms for generating commitments, PSAs can take certain steps to increase the audience's commitment to the issue. One possibility is to encourage the audience to contact the sponsoring organization for a bumper sticker, refrigerator magnet, or t-shirt that declares support for the campaign. Not only does this get the target audience more involved in the issue by phoning or writing for these products, but wearing a t-shirt or displaying a magnet or bumper sticker provides a public display of association and at the same time commitment to the issue. From here, these individuals could be seen as interpersonal communicators who are definitely likely to internalize their support for the issue. It is also likely that the bumper sticker, magnet, or t-shirt could be a conversation piece, in which case the target's friends or neighbors might learn more about the issue or at least gain normative information about how others feel about the issue. In this way providing access to such paraphernalia should enhance the targets' involvement in and commitment to the issue, while at the same time spreading the information along an interpersonal route.

In this issue McKenzie-Mohr provides an example of the use of a decal to increase Canadian residents' composting. Those who were already composting at the time of the study were phoned and asked to place a decal indicating their support of composting on the side of their garbage container. Over 80% of those contacted agreed to this request. It's likely that the visibility of these decals increased the residents' commitment to composting and also informed neighbors of the prevalence of this behavior.

Overall Guidelines for Proenvironmental PSAs

This article provides numerous suggestions for improving proenvironmental PSAs. Several researchers have already developed models that can be readily applied to this goal, and three seem especially suitable for our purposes: McGuire's (1989) input/output matrix, Petty and Cacioppo's (1981) elaboration-likelihood model, and the checklist constructed by Atkin and Freimuth (1989).

During the preproduction stage of the campaign design, developers should focus on the target population. They should first identify the target audience by examining who is most at risk for performing the undesirable behavior and who is most open to media persuasion. The designers should also determine which behaviors are most amenable to change via PSAs. Once a target audience has been identified, members of that audience should be surveyed to learn what experiences they have that are relevant to the issue. The most important criterion for the spokesperson chosen to present the message is that he or she be credible.

During the concept development phase, the campaign designers should begin the first stage of pretesting. Atkin and Freimuth (1989) recommended using focus group interviews or theater testing to present a sample audience with partially formulated ideas. The audience should be asked to respond to such questions as: Did you find the wording appropriate? Did the message explain how to resolve a specific problem? Did you find it personally useful? How involved did you get in the message? Did the message relate to your personal life? Was it well-organized? Were the main arguments vivid? Was background information vivid? and Did you get a sense of norms regarding the issue? In addition the audience should respond to McGuire's output matrix: Did you attend to the message? Did you like it? Were you interested in it? Did you understand it or learn what it was encouraging? Did you learn how to react? Did your attitude change? and Did you agree with it? A posttest would be an effective way to gain information further down the matrix and learn if the audience remembered the message; if they could retrieve the information from memory; whether they had behaved in accord with the message; whether they were reinforced for this; and whether there had been any postbehavioral consolidation of audience members' self-concepts.

During the pretesting at message execution phase, completed messages are presented to a sample audience in preliminary form. Many of the questions presented during the concept development phase will be repeated here. The developers will attempt to learn if the message had strong attentional value; whether the audience comprehended it and found it relevant; what the strengths and weaknesses of the message were; whether it demonstrated how to resolve a specific goal; whether it was useful; whether the audience got involved in the message content; whether it related to the audience's personal life; whether it was vivid; whether it demonstrated norms; and whether the audience felt a sense of commitment to the issue.

In addition to this three-step guideline, there are some general suggestions for campaign designers. The message content should be very specific. As Pratkanis and Greenwald (1993) found, messages that describe how to solve specific problems are more likely to be attended to, as they have been found to break through the message-dense environment. The message should explain precisely how a behavior change should occur, and this explanation should be vivid and involving without having vivid and distracting additional information. The message should include an encoding cue (Keller, 1987) that will definitely be present in the upcoming behavioral setting. This cue should be tied to the main point of the message so that motivation is activated perhaps by providing a descriptive and/or an injunctive norm (Cialdini, Reno, & Kallgren, 1990). This cue will increase the likelihood that the audience will recall the message and act in accord with it at the appropriate point in time.

Atkin and Freimuth provide an excellent checklist for evaluating an initial campaign plan, and combined with McGuire's input/output axis the two provide a thorough guideline for proenvironmental communication developers. If the campaign developer tests the message using these checklists and finds that the message is effective at each stage of the checklists, then not only will the central route to persuasion be incorporated, but the message is much more likely to be effective, which is the most desirable quality of a proenvironmental (or any other) PSA.

References

Atkin, C. K., & Freimuth, V. (1989). Formative evaluation research in campaign design. In R. Rice & C. Atkin (Eds.), *Public communication campaigns* (pp. 131–150). CA: Sage.

Bator, R. J. (1997). *Effective public service announcements: Linking social norms to visual memory cues*. Unpublished doctoral dissertation, Arizona State University, Tempe.

Baumgardner, M. H., Leippe, M. R., Ronis, D. L., & Greenwald, A. G. (1983). In search of reliable persuasion effects: II. Associative interference and persistence in persuasion in a message dense environment. *Journal of Personality and Social Psychology, 45*, 524–537.

Bem, D. J. (1972). Self-perception theory. In L. Berkowitz (Ed.), *Advances in experimental social psychology* (Vol. 6, pp. 1–162). New York: Academic.

Bertrand, J. T. (1978). *Communications pretesting*. Chicago: University of Chicago, Community and Family Study Center.

Chaffee, S. H., & McLeod, J. M. (1973). Consumer decisions and information use. In S. Ward & T. S. Robertson (Eds.), *Consumer behavior: Theoretical sources* (pp. 385–415). Englewood Cliffs, NJ: Prentice-Hall.

Chaiken, S. (1980). Heuristic versus systematic information processing and the use of source versus message cues in persuasion. *Journal of Personality and Social Psychology, 39*, 752–766.

Cialdini, R. B. (1993). *Influence: Science and practice* (3rd ed.). New York: HarperCollins.

Cialdini, R. B., Kallgren, C. A., & Reno, R. R. (1991). A focus theory of normative conduct: A theoretical refinement and reevaluation of the role of norms in human behavior. *Advances in Experimental Social Psychology, 24*, 201–234.

Cialdini, R. B., Reno, R. R., & Kallgren, C. A. (1990). A focus theory of normative conduct: Recycling the concept of norms to reduce littering in public places. *Journal of Personality and Social Psychology, 58*, 1015–1026.

Condelli, L., Archer, D., Aronson, E., Curbow, B., McLeod, B., Pettigrew, T. F., White, L. W., & Yates, S. (1984). Improving utility conservation programs: Outcomes, interventions, and evaluations. *Energy, 9*, 485–494.

Cook, T. D., & Flay, B. R. (1978). The persistence of experimentally induced attitude change. In L. Berkowitz (Ed.), *Advances in experimental social psychology* (Vol. 11, pp. 1–57). San Diego, CA: Academic Press.

Craik, F. I. M. (1981). Encoding and retreival effects in human memory: A partial review. In A. D. Baddeley and J. Long (Eds.), *Attention and performance* (Vol. 9, pp. 110–134). Hillsdale, NJ: Lawrence Erlbaum.

Dwyer, K. J. (1999, January 6). Impact of "Iron Eyes" recalled locally. *Jacksonville Daily News* [On-line]. Available: http://www.jacksonvilledailynews.com/stories/1999/01/06/news08. shtml

Festinger, L. (1957). *A theory of cognitive dissonance.* Stanford, CA: Stanford University Press.

Fisher, R. P., & Craik, F. I. M. (1977). Interaction between encoding and retrieval operations in cued recall. *Journal of Experimental Psychology: Human Learning & Memory, 3*, 701–711.

Flay, B. R., & Best, J. (1982). Overcoming design problems in evaluating health behavior programs. *Evaluation and the Health Professions, 5*, 43–69.

Flay, B. R., & Cook, T. D. (1989). Three models for summative evaluation of prevention campaigns with a mass media component. In R. Rice & C. Atkin (Eds.), *Public communication campaigns* (pp. 175–195). Newbury Park, CA: Sage.

Heider, F. (1946). Attitudes and cognitive organization. *Journal of Psychology, 21*, 107–112.

Hornik, R. C. (1989). Channel effectiveness in development communication programs. In R. Rice & C. Atkin (Eds.), *Public communication campaigns* (pp. 309–330). Newbury Park, CA: Sage.

Hovland, C. I., & Weiss, W. (1951). The influence of source credibility on communication effectiveness. *Public Opinion Quarterly, 15*, 635–650.

Katzev, R., & Johnson, T. (1984). A social-psychological analysis of residential electricity consumption: The impact of minimal justification techniques. *Journal of Economic Psychology, 3*, 267–284.

Keller, K. L. (1987). Memory factors in advertising: The effect of advertising retrieval cues on brand evaluations. *Journal of Consumer Research, 14*, 316–333.

Kelman, H. C., & Hovland, C. I. (1953). "Reinstatement" of the communicator in delayed measurement of opinion change. *Journal of Abnormal and Social Psychology, 48*, 327–335.

Leonhardt, D. (1998, November 9). Got milked? After a $385 million campaign, sales are declining. *Business Week*, p. 52.

McGuire, W. J. (1989). Theoretical foundations of campaigns. In R. Rice & C. Atkin (Eds.), *Public communication campaigns* (pp. 43–65). Newberry Park, CA: Sage.

Mendelsohn, H. (1973). Some reasons why information campaigns can succeed. *Public Opinion Quarterly, 37*, 50–61.

Nisbett, R., & Ross, L. (1980). *Human inference: Strategies and shortcomings of social judgment.* Englewood Cliffs: NJ: Prentice-Hall.

Pallak, M. S., Cook, D. A., & Sullivan, J. J. (1980). Commitment and energy conservation. *Applied Social Psychology Annual, 1*, 235–253.

Petty, R., & Cacioppo, J. (1981). *Attitudes and persuasion: Classic and contemporary approaches.* Dubuque, IA: Brown.

Petty, R., & Cacioppo, J. (1986). *Communication and persuasion: Central and peripheral routes to attitude change.* New York: Springer-Verlag.

Pratkanis, A. R., & Greenwald, A. G. (1993). Consumer involvement, message attention, and the persistence of persuasive impact in a message-dense environment. *Psychology & Marketing, 10*, 321–332.

Rhoads, K. (1994). *The impact of figural vividness on persuasion.* Unpublished master's thesis, Arizona State University, Tempe.

Rice, R. E., & Atkin, C. K. (Eds.). (1989). *Public communication campaigns* (2nd ed.). Newbury Park, CA: Sage.

Rogers, E. M., & Storey, D. (1987). Communication campaigns. In C. Berger & S. Chaffee (Eds.), *Handbook of communication science* (pp. 817–846). Newbury Park, CA: Sage.

Tulving, E. (1979). Relation between encoding specificity and levels of processing. In L. S. Cermak and F. I. M. Craik (Eds.), *Levels of processing in human memory*. Hillsdale, NJ: Lawrence Erlbaum.

Tulving, E., & Osler, S. (1968). Effectiveness of retrieval cues in memory for words. *Journal of Experimental Psychology, 77*, 593–601.

U.S. Department of Health and Human Services. (1984). *Pretesting in health communications* (NIH Publication No. 84-1493). Bethesda, MD: National Cancer Institute.

Webb, P. H., & Ray, M. L. (1979). Effects of TV clutter. *Journal of Advertising Research, 19*, 7–12.

Wood, W. (1982). Retrieval of attitude-relevant information from memory: Effects on susceptibility to persuasion and on intrinsic motivation. *Journal of Personality and Social Psychology, 42*, 798–810.

Wood, W., Kallgren, C. A., & Preisler, R. M. (1985). Access to attitude-relevant information in memory as a determinant of persuasion: The role of message attributes. *Journal of Experimental Social Psychology, 21*, 73–85.

RENEE BATOR earned her undergraduate degree in Psychology at the University of California at Santa Cruz. She earned her master's and doctoral degrees from the Social Psychology program at Arizona State University. She is an Assistant Professor in the Psychology Department at the State University of New York at Plattsburgh, where she teaches classes in social psychology, research methods, and statistics. Her research interests include the application of persuasion theory to prosocial outcomes.

ROBERT CIALDINI received undergraduate, graduate, and postgraduate education in Psychology at the University of Wisconsin, the University of North Carolina, and Columbia University, respectively. He has held Visiting Scholar appointments at Ohio State University, the Universities of California at San Diego and Santa Cruz, the Annenberg School of Communications, and both the Psychology Department and the Graduate School of Business of Stanford University. He is currently Regents' Professor of Psychology at Arizona State University, where he has also been named Distinguished Graduate Research Professor. He has been elected president of the Society of Personality and Social Psychology and of the Personality and Social Psychology Division of the American Psychological Association.

Tulving, E. (1979). Relation between encoding specificity and levels of processing. In L. S. Cermak and F. I. M. Craik (Eds.), *Levels of processing in human memory*. Hillsdale, NJ: Lawrence Erlbaum.

Tulving, E., & Osler, S. (1968). Effectiveness of retrieval cues in memory for words. *Journal of Experimental Psychology*, *77*, 593–601.

U.S. Department of Health and Human Services. (1984). *Pretesting in health communications* (NIH Publication No. 84-1493). Bethesda, MD: National Cancer Institute.

Webb, P. H., & Ray, M. L. (1979). Effects of TV clutter. *Journal of Advertising Research*, **19**, 7–12.

Wood, W. (1982). Retrieval of attitude-relevant information from memory: Effects on susceptibility to persuasion and on intrinsic motivation. *Journal of Personality and Social Psychology*, *42*, 798–810.

Wood, W., Kallgren, C. A., & Preisler, R. M. (1985). Access to attitude-relevant information in memory as a determinant of persuasion: The role of message attributes. *Journal of Experimental Social Psychology*, *21*, 73–85.

RENEE BATOR earned her undergraduate degree in Psychology at the University of California at Santa Cruz. She earned her master's and doctoral degrees from the Social Psychology program at Arizona State University. She is an Assistant Professor in the Psychology Department at the State University of New York at Plattsburgh where she teaches classes in social psychology, research methods, and statistics. Her research interests include the application of persuasion theory to prosocial issues.

ROBERT B. CIALDINI received undergraduate, graduate, and postgraduate education in Psychology at the University of Wisconsin, the University of North Carolina, and Columbia University, respectively. He has held Visiting Scholar appointments at Ohio State University, the Universities of California at San Diego and Santa Cruz, the Annenberg School of Communications, and both the Psychology Department and the Graduate School of Business of Stanford University. He is currently Regents' Professor of Psychology at Arizona State University where he has also been named Distinguished Graduate Research Professor. He has been elected president of both the Society of Personality and Social Psychology and of the Personality and Social Psychology Division of the American Psychological Association.

Journal of Social Issues, Vol. 56, No. 3, 2000, pp. 543–554

Promoting Sustainable Behavior: An Introduction to Community-Based Social Marketing

Doug McKenzie-Mohr*

St. Thomas University

Most programs to foster sustainable behavior continue to be based upon models of behavior change that psychological research has found to be limited. Although psychology has much to contribute to the design of effective programs to foster sustainable behavior, little attention has been paid to ensuring that psychological knowledge is accessible to those who design environmental programs. This article presents a process, community-based social marketing, that attempts to make psychological knowledge relevant and accessible to these individuals. Further, it provides two case studies in which program planners have utilized this approach to deliver their initiatives. Finally, it reflects on the obstacles that exist to incorporating psychological expertise into programs to promote sustainable behavior.

> *Don't let us forget that the causes of human actions are usually immeasurably more complex than our subsequent explanations of them.*
>
> —Fyodor Dostoevsky

I have a simple wish. Each time I journey to the library to review new contributions to the environmental psychology literature, I hope that I will see an individual whom I know, from either a nongovernmental organization, or the Department of the Environment, or the city, who works on environmental programs. My wish is that I will find this individual reviewing the literature and contemplating how best to apply it to program delivery. I have carried this wish for a decade now and it is yet to be realized. Consequently, I have become increasingly convinced that despite our desire to contribute to the attainment of a sustainable future, our publications contribute far more to career advancement than they do to environmental

---

*Correspondence regarding this article should be addressed to Doug McKenzie-Mohr, Department of Psychology, St. Thomas University, Fredericton, New Brunswick, Canada E3B 5G3 [e-mail: McKenzie@StThomasU.ca].

betterment. We have created a psychological literature that is largely invisible to those who can most benefit from it. Lack of visibility, however, does not equal irrelevance. Changing individual behavior is central to achieving a sustainable future. Accordingly, psychology is of considerable relevance to the delivery of effective environmental programs. Desirable goals, such as lowering greenhouse gas emissions, reducing waste, and increasing energy and water efficiency can be met only if high levels of public participation are achieved. Despite the apparent importance of psychological knowledge to effective program design, program planners have yet to widely access or utilize it. Indeed, my experience in working with these individuals has led me to believe that most are not aware that our literature exists or of its relevance to their efforts. I expect that the pressures that exist to publish in academic journals has led to few attempts to make our expertise accessible to those who can most benefit from it. Until we do this we can feel self-righteous in conducting environmental research, but I doubt that we are participating in a truly meaningful enterprise. In short, until we reach out to the individuals who design and deliver environmental programs, our efforts will remain invisible to those who can most benefit from them.

This article presents one attempt to make psychological knowledge visible and relevant to program planners. It outlines a process, community-based social marketing, for developing and delivering environmental programs that is based on psychological expertise. This process has now been presented via workshops, publications, and a Web site (www.cbsm.com) to several thousand program planners in Canada (Kassirer & McKenzie-Mohr, 1998; McKenzie-Mohr, 1996; McKenzie-Mohr & Smith, 1999). This article also presents two attempts by planners to apply this information and reflects on the challenges faced in its application.

To date, most programs to foster sustainable behavior have been information-intensive. In these campaigns, media advertising and the distribution of printed materials are used to foster behavior change. Information-intensive campaigns are usually based on one of two perspectives on behavior change. With the first, program planners assume that by enhancing knowledge of an issue, such as global warming, and encouraging the development of attitudes that are supportive of an activity, such as using mass transit, behavior will change. Unfortunately, a variety of studies have established that enhancing knowledge and creating supportive attitudes often has little or no impact upon behavior. For example:

- Householders who were interested in enhancing the energy efficiency of their homes participated in a comprehensive workshop on residential energy conservation. Despite significant changes in knowledge and attitudes, behavior did not change (Geller, 1981).

- Householders who volunteered to participate in a 10-week study of water conservation received a booklet that described the relationship between water use and energy use, and methods were described that could conserve water. Even though great attention was given to preparing the booklet, it had no impact upon water consumption (Geller, Erickson, & Buttram, 1983).

- Two surveys of Swiss respondents found that environmental attitudes and knowledge were poorly associated with environmental behavior (Finger, 1994).

- When 500 people were interviewed regarding their personal responsibility for picking up litter, 94% acknowledged responsibility. When leaving the interview, however, only 2% picked up litter that had been "planted" by the researcher (Bickman, 1972).

The second perspective suggests that behavior is strongly influenced by economic motives. When planners adopt this perspective, they are apt to deliver programs that highlight the economic advantages of engaging in a specific activity, such as installing compact fluorescent bulbs, assuming that the public is "rational" and will act in their economic self-interest. As before, information-intensive programs that have been based on this perspective have also been largely unsuccessful. For instance, California utilities annually spend $200 million to foster residential energy efficiency through the purchase of energy-efficient innovations, such as programmable thermostats, or through lifestyle changes, such as turning down air conditioning before leaving for work (Costanzo, Archer, Aronson, & Pettigrew, 1986). Despite this expensive advertising campaign, household energy use has remained essentially unaltered. Similarly, when the Residential Conservation Service (RCS) was brought into existence by an act of the U.S. Congress in 1978, utilities were mandated to provide their customers with free home energy audits, low-cost loans, and information on contractors and suppliers. Evaluations of this effort suggest that on average energy use per household was reduced by 2–3% (Hirst, 1984; Hirst, Berry, & Soderstrom, 1981; U.S. Department of Energy, 1984). Considering that millions of dollars were spent on the RCS and that energy savings of substantially more than 2–3% are attainable, this initiative can only be viewed as a failure. A U.S. National Research Council report concluded that the RCS overlooks "the rich mixture of cultural practices, social interactions, and human feelings that influence the behavior of individuals, social groups and institutions" (Stern & Aronson, 1984).

Information campaigns likely proliferate because it is comparably easy to air radio or television advertisements or distribute printed material. Advertising, however, is often a very expensive way of reaching people. In one extreme case, a California utility spent more money on advertising the benefits of insulation than it would have cost to upgrade the insulation of targeted homes (Pope, 1982). The failure of mass-media campaigns to foster sustainable behavior is due to some extent to inadequate design of the messages, but more importantly to an underestimation of the difficulty of changing behavior (Costanzo et al., 1986). Costanzo et al. note that most mass-media campaigns to promote energy efficiency are based upon traditional marketing techniques in which the sustainable activity is viewed as a "product" to be sold. Advertising, they indicate, is effective in altering our preference to purchase one brand over another. Altering consumer preferences, however, is not creating new behavior. As they note: "These small changes in behavior generally require little expense or effort and no dramatic change in lifestyle" (p. 256). In contrast, promoting engagement in a new activity, such as walking or biking to work, is much more complex. An array of barriers to these activities exist, such as concerns over time, safety, weather, and convenience. The diversity of barriers that exist for any sustainable activity means that information campaigns alone will rarely bring about behavior change.

In Canada, community-based social marketing has emerged as an attractive alternative to information-intensive campaigns. This emergence can be traced to a growing understanding on the part of program planners that conventional campaigns, which rely heavily or exclusively on media advertising, can be effective in creating public awareness and improved understanding of issues but are limited in their ability to foster behavior change (Aronson & Gonzales, 1990; Costanzo et al., 1986; Yates & Aronson, 1983).

Community-Based Social Marketing

Community-based social marketing is composed of four steps: uncovering barriers to behaviors and then, based upon this information, selecting which behavior to promote; designing a program to overcome the barriers to the selected behavior; piloting the program; and then evaluating it once it is broadly implemented (McKenzie-Mohr & Smith, 1999). Community-based social marketing merges knowledge from psychology with expertise from social marketing (see also Geller, 1989). Social marketing emphasizes that effective program design begins with understanding the barriers people perceive to engaging in an activity (see, for example, Andreasen, 1995). Social marketing also underscores the importance of strategically delivering programs so that they target specific segments of the public and overcome the barriers to this segment's engaging in the behavior.

Uncovering Barriers and Selecting Behaviors

Reduction of the municipal solid waste stream can occur from a variety of activities, such as recycling, source reduction, or reuse. Similarly, lowering greenhouse gas emissions can be achieved by such actions as using alternative transportation (carpooling, bicycling, telecommuting) or lowering household energy use (upgrading insulation levels, installing low-flow showerheads, or closing blinds before leaving for work). Although it might be desirable to promote all of these behaviors, resources rarely exist to foster public participation in a wide range of activities. Consequently, it is necessary to make an informed decision regarding which behavior(s) to promote. With community-based social marketing, the decision regarding which behavior(s) to promote is based primarily upon the answer to three questions. First, what is the potential impact of the behavior? That is, what level of reduction in greenhouse gases are achievable, for example, through modal transportation shifts or the purchase of more-energy-efficient vehicles? Second, what barriers exist to engaging in these activities? In deciding which behavior to promote, it is important to know what the barriers are to broad public participation in the activity. In a limited number of cases, the psychological literature has already identified barriers (see, for example, McKenzie-Mohr, Nemiroff, Beers, & Desmarais, 1995; Schultz, Oskamp, & Mainieri, 1995), though frequently this information needs to be contextualized. For instance, in colder climates winter can be a significant barrier to year-round backyard composting, whereas weather may not be a consideration at all in other areas. In many cases, barriers have not been identified (see Stern & Oskamp, 1987, for a review of the environmental psychology literature), necessitating that preliminary research be conducted prior to deciding which behavior(s) to promote. In identifying barriers, social marketers often identify differences between individuals who engage in the activity and those who do not. Several research methods can be utilized to uncover these differences, including focus groups, observational studies, and survey research. Further, statistical techniques, such as discriminant analysis and logistic regression, can be particularly useful in identifying and prioritizing differences. For example, these techniques were used to distinguish householders who engage in backyard composting from those who did not (McKenzie-Mohr et al., 1995). This research revealed that in comparison to noncomposters, individuals who compost perceive reducing waste as being more important, and composting as less unpleasant, inconvenient, and time consuming.

Barriers to a behavior may be either internal (e.g., lacking the perceived skill to install a programmable thermostat) or external (e.g., absence of programmable thermostats locally; see Stern, this issue). Also, numerous barriers exist for any behavior, and these barriers appear to be behavior specific (McKenzie-Mohr et al., 1995; Oskamp, 1995; Tracy & Oskamp, 1983–84). That is, what impedes an individual, for example, from walking to work is distinct from what might preclude her from closing the blinds each morning or purchasing products with recycled content. Accordingly, the genesis of a sound community-based social marketing strategy is identifying barriers. Without detailed knowledge of barriers, it is highly unlikely that an effective strategy can be developed. Psychological expertise in research methods and statistical techniques can contribute significantly to the uncovering of barriers and the development of sound strategies.

The third question to be asked in determining which behavior(s) to promote is whether the resources exist to overcome identified barriers. An important consideration in contemplating the answer to this question is whether the behavior is one-time (e.g., purchasing an energy efficient vehicle) or repetitive (e.g., closing blinds each day before leaving for work). In general, it is more difficult to alter and maintain repetitive behavior changes than it is to bring about one-time changes in behavior (see for example, Kempton, Darley, & Stern, 1992; Kempton, Harris, Keith, & Weihl, 1984).

Designing Strategies

An effective social marketing strategy removes barriers to the behavior to be promoted. For example, in fostering the purchase of products with recycled content, the King County Commission in Washington State first identified barriers to their purchase and then systematically removed them (Herrick, 1995). Survey and focus group research indicated the existence of five barriers to the purchase of these products. The commission felt that little could be done with respect to two of these barriers: the perception that these products cost more and were of inferior quality. The three other barriers, low awareness of which products had recycled content, suspicion regarding environmental claims of manufacturers, and the diffi- culty of quickly identifying these products while shopping, could, however, be overcome. Although this program utilized traditional media and in-store advertis- ing, it relied primarily upon a shelf prompt that advertised that a product had recy- cled content. The results from this social marketing strategy demonstrate the importance of first identifying barriers and then systematically removing them. Analysis of electronic inventories of participating retail stores indicated that purchases of recycled-content products rose 27% as a consequence of this social marketing strategy. This successful program has now been adopted by a number of cities throughout the United States.

Psychological expertise can be readily applied to removing barriers to behav- ior change (see McKenzie-Mohr & Smith, 1999, for an overview of how this knowledge can be applied to program design). For example, when low motivation exists to engage in a sustainable behavior, it can be enhanced through the use of commitment strategies (see Katzev & Wang, 1994) or incentives (see Gardner & Stern, 1996). When individuals do not perceive an activity as being the "right thing to do," knowledge regarding the use of injunctive and descriptive norms can be applied (see, for example, Cialdini, Reno, & Kallgren, 1990). Numerous other applications of psychological knowledge to strategy design can be made (see Bator & Cialdini, this issue, for further examples).

Piloting

Following the development of a strategy, it should be piloted prior to being broadly implemented. Once again, psychological expertise in research methods and statistics can lead to cost-effective and definitive pilots. With community-based social marketing, pilots are repeated until the desired level of behavior change has been achieved.

Evaluation

Despite the expense of delivering many environmental programs, evaluations of their effectiveness are infrequent. Community-based social marketing stresses the evaluation of implemented programs. Further, it emphasizes the direct measurement of behavior or its consequences (e.g., energy use) rather than relying on self-report measures.

Case Studies

Community-based social marketing has now been applied in a variety of projects across Canada. Here are two examples (others can be found at www.cbsm.com and www.toolsofchange.com).

Backyard Composting

The province of Nova Scotia recently announced a ban of all organic materials from landfills. In response, municipalities throughout the province are developing initiatives to remove organics from the waste stream. In King and Annapolis County, local officials decided to promote backyard composting as their preferred method of meeting this ban. Following the principles of community-based social marketing, they first conducted survey research to identify local barriers to backyard composting and determine present levels of backyard composting. This research identified that a surprisingly high number of residents (56%) were composting. Further, this research indicated that in comparison with composters, those who were not composting perceived it to be inconvenient and unpleasant, not the "right thing to do," and lacked basic knowledge on how to compost. Based on a review of the psychological literature, the program planners developed a unique initiative to leverage current levels of composting and overcome identified barriers. Given the high number of householders who were already composting, it was decided to leverage this participation in encouraging others to backyard compost. Students contacted local residents by telephone and asked them if they presently composted. Those who did were asked to make two commitments. The program planners reasoned that one explanation for the absence of community norms supporting backyard composting was the relative invisibility of composting compared to other activities, such as curbside recycling. Accordingly, those who composted were asked to commit to placing a decal on the side of their blue box or garbage container indicating that they composted. As a form of commitment, the act of placing a decal on the side of their blue box or garbage container served to increase the likelihood that the household would compost more effectively, while at the same time fostering the development of descriptive social norms (Cialdini et al., 1990) in which composting is seen as appropriate behavior.

Based upon previous research that has successfully utilized commitments to spread the adoption of a new technique, grass cycling (Cobern, Porter, Leeming, & Dwyer, 1995), a similar approach was used in this project. Householders who composted were asked to speak to their neighbors about composting and provide them with a package that dispelled perceptions that it was unpleasant and inconvenient, and provided requisite information on how to compost. While fully 81% agreed to place a decal on their blue box or garbage container, very few were willing to speak to their neighbors. This reluctance was a significant setback to the delivery of the program and underscores the importance of piloting strategies before broad implementation.

Those who indicated on the telephone that they did not compost were asked if they would be interested in beginning to compost. Those who expressed interest were visited by an employee who addressed the specific barriers that had been identified in the survey research. Although funding did not allow evaluation of this project, a pilot project that had been conducted the previous year, upon which this larger project was partially based, revealed that 80% of those household residents who had expressed an interest in composting were found to be composting in a follow-up several months later (K. Donnelly, personal communication, 1999).

Encouraging Water Efficiency

As a consequence of lawn watering, summer water use can rise 50% relative to other times of the year. In an effort to offset the cost of building a new water-processing plant, Durham Region, Ontario, developed a community-based social marketing strategy to reduce water use by 10% (Durham Region, 1997). Through survey techniques and direct observation, barriers to water-efficient lawn care were identified. Pilot households were divided into two groups. Householders in the first group were visited by a student employee on bicycle who spoke to residents about efficient water use. Although psychological knowledge was not used to shape the presentation of this information, residents were provided with a water gauge (one identified barrier was that residents were unaware of when they had watered their lawn adequately) and a prompt that was to be placed over the outside water faucet that reminded residents to water their lawn on either odd or even calendar days based upon their house numbers and to water their lawns only when it had not rained in the previous week. Further, these residents were asked to sign commitments that they would water their lawns only on odd or even days and that they would limit their watering to one inch per week (72% of those approached made these commitments). Meanwhile, those householders who were in the "information only" condition were provided with an information packet on efficient water use. Compared to baseline measurements, observation of residents indicated that those householders who were visited by cyclists decreased watering by 54%, whereas those in the "information only" control group increased lawn watering by 15%. Further, watering lawns for longer than 1 hour decreased by 66% when householders were visited by a cyclist, whereas it increased by 96% in the other condition. In total, this program cost $88 (Canadian) to deliver per household, for a total program cost of $80,000. Durham Region calculates that the achieved reduction in peak water consumption allowed 250 new homes to be serviced with a savings in water plant development costs of $945,000.

Reflections

In the experience of the author, psychologists are most likely to have an influence in the area of program design. That is, program planners are receptive to techniques that they can employ easily, such as the use of commitment strategies or vivid communications, into the delivery of their programs (see Bator & Cialdini, this issue, for further examples). The other components of community-based social marketing—identifying behaviors and their barriers, piloting, and evaluation—are far less likely to be utilized. It is useful to reflect on why program planners are less likely to incorporate other central aspects of sound program design and delivery.

Although identifying barriers is a critical step in deciding whether it is wise to attempt to promote a specific behavior as well as craft a social marketing strategy, significant pressures exist to skip this step. Indeed, in a recent review of Canadian environmental programs, most programs were found not to identify barriers prior to developing strategies (Kassirer & McKenzie-Mohr, 1998). A variety of reasons exist for not identifying barriers. Three of the most common include

- Program planners are likely to believe that the barriers to an activity are already well known.
- Most programs must be delivered within a short period of time, which makes conducting barrier research a challenge.
- The organizations that deliver these programs suffer from financial constraints that make additional work difficult to justify.

Social psychological research suggests that we readily form personal theories regarding the behavior of others and then search selectively for information that confirms our beliefs. This suggests that program planners are apt to believe that they already fully understand the barriers to an activity, independent of whether they actually do. Although as psychologists we may not be able easily to persuade them that their personal theories may be in error (particularly when programs are not evaluated and, therefore, do not provide feedback on their efficacy to their designer), we are more likely to be effective if we can provide research findings on the barriers to an activity that they are interested in promoting. To date, psychological research on barriers primarily has been confined to energy efficiency and waste reduction. We quickly need to develop knowledge regarding the barriers to a much broader set of activities. Further, we need to participate in interdisciplinary efforts to identify the most important activities to research.

Conducting barrier research will add significantly to the length of time required to deliver a project. In many cases, it is reasonable to assume that collecting this information can add 4 to 8 weeks to the length of a project. Further, obtaining this information can add substantially to the cost of delivering a program. This additional time and cost are likely to pale, however, compared to the time and cost of redelivering a program because the first attempt failed to change behavior. It would be useful if we could provide return-on-investment (ROI) information that compared the relative success of projects in which barriers were first identified with those in which they were not.

As with identifying barriers, time and financial constraints also limit the likelihood that programs will be piloted or evaluated. Given that psychological research has revealed that many programs do not change behavior, adopting pilots and evaluations is particularly important.

Over the last several years, I have been attempting to make psychological knowledge more accessible to program planners through delivering workshops and writing specifically for them and by developing a Web site (www.cbsm.com) that allows easier access to relevant information. For example, the Web site provides a guide to fostering sustainable behavior and searchable databases of relevant articles, case studies, and graphics. Further, the site provides the opportunity for program planners to share information with one another and with psychologists through a discussion forum. The feedback that I have received on these attempts to make psychological knowledge more visible suggests that program planners are willing recipients of this information and are anxious to have a dialogue with psychologists regarding program delivery. To ensure that this happens, we need to make certain that attempts by psychologists to work more actively with program planners are not an impediment to tenure and promotion.

Conclusion

To date, little attention has been paid to ensuring that psychological expertise regarding behavior change in general, and fostering sustainable behavior in particular, is shared with program planners. Substantial opportunities exist to work with these individuals in promoting a wide range of sustainable behaviors. As environmental psychologists we need to consider how best to share our expertise with program planners and ensure that our efforts are well integrated with their needs. Behavior change may be central to the transition to a sustainable future, but psychological knowledge has yet to become central to the development of initiatives to foster sustainable behavior.

References

Andreasen, A. (1995). *Marketing social change: Changing behavior to promote health, social development, and the environment.* San Francisco: Jossey-Bass.

Aronson, E., & Gonzales, M. H. (1990). Alternative social influence processes applied to energy conservation. In J. Edwards, R. S. Tindale, L. Heath, & E. J. Posaval (Eds.), *Social influences: Processes and prevention* (pp. 301–325). New York: Plenum.

Bickman, L. (1972). Environmental attitudes and actions. *Journal of Social Psychology, 87,* 323–324.

Cialdini, R. B., Reno, R. R., & Kallgren, C. A. (1990). A focus theory of normative conduct: Recycling the concept of norms to reduce littering in public places. *Journal of Personality and Social Psychology, 58,* 1015–1026.

Cobern, M. K., Porter, B. E., Leeming, F. C., & Dwyer, W. O. (1995). The effect of commitment on adoption and diffusion of grass cycling. *Environment and Behavior, 27,* 213–232.

Costanzo, M., Archer, D., Aronson, E., & Pettigrew, T. (1986). Energy conservation behavior: The difficult path from information to action. *American Psychologist, 41,* 521–528.

Durham Region. (1997). *Durham Region Outdoor Water Conservation Pilot Study.* Durham Region, Ontario, Canada: Author.

Finger, M. (1994). From knowledge to action? Exploring the relationships between environmental experiences, learning, and behavior. *Journal of Social Issues, 50*(3), 141–160.

Gardner, G. T., & Stern, P. C. (1996). *Environmental problems and human behavior.* Boston: Allyn and Bacon.

Geller, E. S. (1981). Evaluating energy conservation programs: Is verbal report enough? *Journal of Consumer Research, 8,* 331–335.

Geller, E. S. (1989). Applied behavior analysis and social marketing: An integration for environmental preservation. *Journal of Social Issues, 45*(1), 17–36.

Geller, E. S., Erickson, J. B., & Buttram, B. A. (1983). Attempts to promote residential water conservation with educational, behavioral and engineering strategies. *Population and Environment Behavioral and Social Issues, 6,* 96–112.

Herrick, D. (1995). Taking it to the stores: Retail sales of recycled products. *Resource Recycling.*

Hirst, E. (1984). Household energy conservation: A review of the federal residential conservation service. *Public Administration Review, 44,* 421–430.

Hirst, E., Berry, L., & Soderstrom, J. (1981). Review of utility home energy audit programs. *Energy, 6,* 621–630.

Kassirer, J., & McKenzie-Mohr, D. (1998). *Tools of change: Proven methods for promoting environmental citizenship.* Ottawa, Ontario, Canada: National Round Table on the Environment and the Economy.

Katzev, R., & Wang, T. (1994). Can commitment change behavior? A case study of environmental actions. *Journal of Social Behavior and Personality, 9,* 13–26.

Kempton, W., Darley, J. M., & Stern, P. C. (1992). Psychological research for the new energy problems: Strategies and opportunities. *American Psychologist, 47,* 1213–1223.

Kempton, W., Harris, C. K., Keith, J. G., & Weihl, J. S. (1984). Do consumers know what works in energy conservation? In J. Harris & C. Blumstein (Eds.), *What works: Documenting energy conservation in buildings* (pp. 429–438). Washington, D.C.: American Council for an Energy Efficient Economy.

McKenzie-Mohr, D. (1996). *Promoting a sustainable future: An introduction to community-based social marketing.* Ottawa, Ontario, Canada: National Round Table on the Environment and the Economy.

McKenzie-Mohr, D., Nemiroff, L. S., Beers, L., & Desmarais, S. (1995). Determinants of responsible environmental behavior, *Journal of Social Issues, 51*(4), 139–156.

McKenzie-Mohr, D., & Smith, W. (1999). *Fostering sustainable behavior: An introduction to community-based social marketing* (2nd ed.). Gabriola Island, British Columbia, Canada: New Society.

Oskamp, S. (1995). Resource conservation and recycling: Behavior and policy. *Journal of Social Issues, 51*(4), 157–177.

Pope, E. (1982, December 10). PG&E's loans aimed at poor miss the mark. *San Jose Mercury*, p. 6B.

Schultz, P. W., Oskamp, S., & Mainieri, T. (1995). Who recycles and when? A review of personal and situational factors. *Journal of Environmental Psychology, 15*, 105–121.

Stern, P. C., & Aronson, E. (Eds.). (1984). *Energy use: The human dimension*. New York: Freeman.

Stern, P. C. & Oskamp, S. (1987). Managing scarce environmental resources. In D. Stokols & I. Altman (Eds.), *Handbook of environmental psychology* (pp. 1043–1088). New York: Wiley.

Tracy, A. P., & Oskamp, S. (1983–84). Relationships among ecologically responsible behaviors. *Journal of Environmental Systems, 13*, 115–126.

U.S. Department of Energy. (1984). *Residential conservation service evaluation report (Hearings before the Committee on Energy and Natural Resources of the United States Senate, Ninety-Eighth Congress)*. Washington, D.C.: U.S. Government Printing Office.

Yates, S. M., & Aronson, E. (1983). A social psychological perspective on energy conservation in residential buildings. *American Psychologist, 38*(4), 435–444.

DOUG MCKENZIE-MOHR is an environmental psychologist teaching at St. Thomas University. He has served as a member of the steering committee of Holis: The Society for a Sustainable Future, the education task force of the Canadian National Round Table on the Environment and the Economy, and the technical advisory group for SustainABILITY. He is a coauthor of *Fostering Sustainable Behavior: An Introduction to Community-Based Social Marketing*.

Journal of Social Issues, Vol. 56, No. 3, 2000, pp. 555–578

Environmental Justice: Grassroots Activism and Its Impact on Public Policy Decision Making

Robert D. Bullard* and Glenn S. Johnson

Clark Atlanta University

A growing body of evidence reveals that people of color and low-income persons have borne greater environmental and health risks than the society at large in their neighborhoods, workplace, and playgrounds. Over the last decade or so, grassroots activists have attempted to change the way government implements environmental, health, and civil rights laws. Grassroots groups have organized, educated, and empowered themselves to improve the way government regulations and environmental policies are administered. A new movement emerged in opposition to environmental racism and environmental injustice. Over the last 2 decades or so, grassroots activists have had some success in changing the way the federal government treats communities of color and their inhabitants. Grassroots groups have also organized, educated, and empowered themselves to improve the way health and environmental policies are administered. Environmentalism is now equated with social justice and civil rights.

Despite significant improvements in environmental protection over the past several decades, millions of Americans continue to live in unsafe and unhealthy physical environments. Many economically impoverished communities and their inhabitants are exposed to greater health hazards in their homes, on the jobs, and in their neighborhoods when compared to their more affluent counterparts (Bryant & Mohai, 1992; Bullard, 1994a).

Hardly a day passes without the media discovering some community or neighborhood fighting a landfill, incinerator, chemical plant, or some other polluting industry. This was not always the case. Just 3 decades ago, the concept of environmental justice had not registered on the radar screens of environmental, civil rights, or social justice groups (Bullard, 1994b). Nevertheless, it should not be forgotten

*Correspondence concerning this article should be addressed to Robert Bullard, Environmental Justice Resource Center, Clark Atlanta University, Atlanta GA 30314 [e-mail: rbullard@cau.edu].

that Martin Luther King, Jr., went to Memphis in 1968 on an environmental and economic justice mission for the striking Black garbage workers. The strikers were demanding equal pay and better work conditions. Of course, Dr. King was assassinated before he could complete his mission.

Another landmark garbage dispute took place a decade later in Houston, when African American homeowners began a bitter fight to keep a sanitary landfill out of their suburban middle-income neighborhood (Bullard, 1983). Residents formed the Northeast Community Action Group (NECAG). NECAG and its attorney, Linda McKeever Bullard, filed a class action lawsuit to block the facility from being built. The 1979 lawsuit, *Bean v. Southwestern Waste Management, Inc.*, was the first of its kind to challenge the siting of a waste facility under civil rights law. The landmark Houston case occurred 3 years before the environmental justice movement was catapulted into the national limelight in rural and mostly African American Warren County, North Carolina.

The environmental justice movement has come a long way since its humble beginning in 1982 in Warren County, North Carolina, where a PCB landfill ignited protests and over 500 arrests. The Warren County protests provided the impetus for a U.S. General Accounting Office (1983) study, *Siting of Hazardous Waste Landfills and Their Correlation With Racial and Economic Status of Surrounding Communities*. That study revealed that three out of four of the off-site commercial hazardous waste landfills in Region 4 (which comprises eight states in the South) happen to be located in predominantly African American communities, although African Americans make up only 20% of the region's population. More important, the protesters put "environmental racism" on the map. Fifteen years later, the state of North Carolina is spending over $25 million to clean up and detoxify the Warren County PCB landfill.

The protests also led the Commission for Racial Justice (1987) to produce *Toxic Wastes and Race in the United States*, the first national study to correlate waste facility sites and demographic characteristics. Race was found to be the most potent variable in predicting where these facilities were located—more powerful than poverty, land values, and home ownership. In 1990, *Dumping in Dixie: Race, Class, and Environmental Quality* chronicled the convergence of two social movements—social justice and environmental movements—into the environmental justice movement (Bullard, 1994a). This book highlighted African Americans' environmental activism in the South, the same region that gave birth to the modern civil rights movement. What started out as local and often isolated community-based struggles against toxics and facility siting blossomed into a multi-issue, multiethnic, and multiregional movement.

The 1991 First National People of Color Environmental Leadership Summit in Washington, D.C., was probably the most important single event in the movement's history. The summit broadened the environmental justice movement beyond its antitoxics focus to include issues of public health, worker safety, land use, transportation, housing, resource allocation, and community empowerment (C. Lee, 1992). The meeting also demonstrated that it is possible to build a multiracial grassroots movement around environmental and economic justice (Alston, 1992).

The four-day summit was attended by over 650 grassroots and national leaders from around the world. Delegates came from all 50 states, including Alaska and Hawaii, as well as from Puerto Rico, Chile, Mexico, and as far away as the Marshall Islands. People attended the summit to share their action strategies, redefine the environmental movement, and develop common plans for addressing environmental problems affecting people of color in the United States and around the world.

On September 27, 1991, summit delegates adopted 17 "principles of environmental justice." These principles were developed as a guide for organizing, networking, and relating to government and nongovernmental organizations (NGOs). By June 1992, Spanish and Portuguese translations of the principles were being used and circulated by NGOs and environmental justice groups at the Earth Summit in Rio de Janeiro.

The publication of the *People of Color Environmental Groups Directory* in 1992 and 1994 further illustrates that environmental justice organizations are found in the United States from coast to coast, in Puerto Rico, in Mexico, and in Canada. Groups have come to embrace a wide range of issues, including children's health, pollution prevention, housing, brownfields (sites that have actual or perceived contamination and may be used as redevelopment sites), community reinvestment, urban sprawl, transportation, land use, and worker safety.

The Environmental Justice Paradigm

Despite significant improvements in environmental protection over the past several decades, millions of Americans continue to live in unsafe and unhealthy physical environments (Institute of Medicine, 1999). Many economically impoverished communities and their inhabitants are exposed to greater health hazards in their homes, on their jobs, and in their neighborhoods when compared to their more affluent counterparts (Bullard, 1994a, 1994b; Bryant, 1995; Bryant & Mohai, 1992; Calloway & Decker, 1997; Collin & Collin, 1998; U.S. EPA, 1992b).

From New York to Los Angeles, grassroots community resistance has emerged in response to practices, policies, and conditions that residents have judged to be unjust, unfair, and illegal. Some of these conditions include (1) unequal enforcement of environmental, civil rights, and public health laws; (2) differential exposure of some populations to harmful chemicals, pesticides, and other toxins in the home, school, neighborhood, and workplace; (3) faulty assumptions in calculating, assessing, and managing risks; (4) discriminatory zoning and land use practices; and (5) exclusionary practices that prevent some individuals and groups from participation in decision making or limit the extent of their participation (Bullard, 1993b; C. Lee, 1992).

Environmental justice is defined as the fair treatment and meaningful involvement of all people regardless of race, color, national origin, or income with respect to the development, implementation, and enforcement of environmental laws, regulations, and policies. Fair treatment means that no group of people, including racial, ethnic, or socioeconomic groups, should bear a disproportionate share of the negative environmental consequences resulting from industrial, municipal, and commercial operations or the execution of federal, state, local, and tribal programs and policies (U.S. EPA, 1998).

During its 30-year history, the U.S. Environmental Protection Agency (EPA) has not always recognized that many government and industry practices (whether intended or unintended) have an adverse impact on poor people and people of color. Growing grassroots community resistance emerged in response to practices, policies, and conditions that residents judged to be unjust, unfair, and illegal. The EPA is mandated to enforce the nation's environmental laws and regulations equally across the board. It is required to protect all Americans, not just individuals or groups who can afford lawyers, lobbyists, and experts. Environmental protection is a right, not a privilege reserved for a few who can "vote with their feet" and escape or fend off environmental stressors.

The current environmental protection apparatus manages, regulates, and distributes risks (Bullard, 1996). The dominant environmental protection paradigm institutionalizes unequal enforcement; trades human health for profit; places the burden of proof on the "victims" and not the polluting industry; legitimates human exposure to harmful chemicals, pesticides, and hazardous substances; promotes "risky" technologies; exploits the vulnerability of economically and politically disenfranchised communities; subsidizes ecological destruction; creates an industry around risk assessment and risk management; delays cleanup actions; and fails to develop pollution prevention as the overarching and dominant strategy (Austin & Schill, 1991; Bullard, 1992, 1993c).

A growing body of evidence reveals that people of color and low-income persons have borne greater environmental and health risks than the society at large in their neighborhoods, workplaces, and playgrounds (Institute of Medicine, 1999; Johnson, Williams, & Harris, 1992; National Institute for Environmental Health Sciences, 1995). On the other hand, the environmental justice paradigm embraces a holistic approach to formulating environmental health policies and regulations; developing risk reduction strategies for multiple, cumulative, and synergistic risks; ensuring public health; enhancing public participation in environmental decision making; promoting community empowerment; building infrastructure for achieving environmental justice and sustainable communities; ensuring interagency cooperation and coordination; developing innovative public-private partnerships and collaboratives; enhancing community-based pollution prevention strategies;

ensuring community-based sustainable economic development; and developing geographically oriented community-wide programming.

The question of environmental justice is not anchored in a debate about whether or not decision makers should tinker with risk assessment and risk management. The environmental justice framework rests on developing tools and strategies to eliminate unfair, unjust, and inequitable conditions and decisions (Bullard, 1996). The framework also attempts to uncover the underlying assumptions that may contribute to and produce differential exposure and unequal protection. It brings to the surface the *ethical* and *political* questions of "who gets what, when, why, and how much." General characteristics of this framework include the following:

- The environmental justice framework adopts a public health model of prevention (i.e., elimination of the threat before harm occurs) as the preferred strategy.

- The environmental justice framework shifts the burden of proof to polluters/dischargers who do harm, who discriminate, or who do not give equal protection to people of color, low-income persons, and other "protected" classes.

- The environmental justice framework allows disparate impact and statistical weight or an "effect" test, as opposed to "intent," to be used to infer discrimination.

- The environmental justice framework redresses disproportionate impact through "targeted" action and resources. In general, this strategy would target resources where environmental and health problems are greatest (as determined by some ranking scheme but not limited to risk assessment).

Dismantling Environmental Racism

In the real world, all communities are not created equal. All communities do not receive equal protection. Economics, political clout, and race play an important part in sorting out residential amenities and disamenities. Racism is alive and well in the United States (Doob, 1993). Environmental racism is as real as the racism found in housing, employment, education, and voting (Bullard, 1993a). *Environmental racism* refers to any environmental policy, practice, or directive that differentially affects or disadvantages (whether intended or unintended) individuals, groups, or communities based on race or color. Environmental racism is one form of environmental injustice and is reinforced by government, legal, economic, political, and military institutions. Environmental racism combines with public policies and industry practices to provide *benefits* for Whites while shifting *costs* to people of color (Bullard, 1993a; Collin, 1992; Colquette & Robertson, 1991; Godsil, 1990).

The impetus behind the environmental justice movement did not come from within government, academia, or largely White, middle-class, nationally based environmental and conservation groups. The impetus for change came from people of color, grassroots activists, and their "bottom-up" leadership approach. Grassroots groups organized themselves, educated themselves, and empowered themselves to make fundamental change in the way environmental protection is administered in their communities.

Government has been slow to ask the questions of who gets help and who does not, who can afford help and who cannot, why some contaminated communities get studied whereas others get left off the research agenda, why industry poisons some communities and not others, why some contaminated communities get cleaned up whereas others do not, why some populations are protected and others are not protected, and why unjust, unfair, and illegal policies and practices are allowed to go unpunished.

Struggles for equal environmental protection and environmental justice did not magically appear in the 1990s. Many communities of color have been engaged in life-and-death struggles for more than a decade. In 1990, the Agency for Toxic Substances and Disease Registry (ATSDR) held a historic conference in Atlanta. The ATSDR National Minority Health Conference focused on contamination (Johnson, Williams, & Harris, 1992). In 1992, after meeting with community leaders, academicians, and civil rights leaders, the EPA (under the leadership of William Reilly) acknowledged there was a problem and established the Office of Environmental Equity (the name was changed to the Office of Environmental Justice under the Clinton administration).

In 1992, the EPA produced one of the first comprehensive documents to examine the whole question of risk, environmental hazards and their equity: *Environmental Equity: Reducing Risk for All Communities* (U.S. EPA, 1992a). The report and the resulting Office of Environmental Equity were initiated only after prodding from people of color, environmental justice leaders, activists, and a few academicians.

In 1993, EPA also established a 25-member National Environmental Justice Advisory Council (NEJAC) under the Federal Advisory Committee Act. The NEJAC is comprised of stakeholders representing grassroots community groups; environmental groups; nongovernmental organizations; state, local, and tribal governments; academia; and industry. The NEJAC divides its environmental justice work into six subcommittees: Health and Research, Waste and Facility Siting, Enforcement, Public Participation and Accountability, Native American and Indigenous Issues, and International Issues.

In February 1994, seven federal agencies, including the ATSDR, the National Institute for Environmental Health Sciences, the EPA, the National Institute of

Occupational Safety and Health, the National Institutes of Health, the Department of Energy (DOE), and the Centers for Disease Control and Prevention sponsored a national health symposium, "Health and Research Needs to Ensure Environmental Justice," in Arlington, Virginia. The conference planning committee was unique in that it included grassroots organization leaders, residents of affected communities, and federal agency representatives. The goal of the February conference was to bring diverse stakeholders and those most affected to the decision-making table (National Institute for Environmental Health Sciences, 1995). Recommendations from the symposium included the following:

- Conduct meaningful health research in support of people of color and low-income communities.
- Promote disease prevention and pollution prevention strategies.
- Promote interagency coordination to ensure environmental justice.
- Provide effective outreach, education, and communications.
- Design legislative and legal remedies.

In response to growing public concern and mounting scientific evidence, President Bill Clinton on February 11, 1994 (the second day of the national health symposium), issued Executive Order 12898, "Federal Actions to Address Environmental Justice in Minority Populations and Low-Income Populations." This order attempts to address environmental injustice within existing federal laws and regulations.

Executive Order 12898 reinforces the 35-year-old Civil Rights Act of 1964, Title VI, which prohibits discriminatory practices in programs receiving federal funds. The order also focuses the spotlight on the National Environmental Policy Act (NEPA), a 25-year-old law that set policy goals for the protection, maintenance, and enhancement of the environment. NEPA's goal is to ensure for all Americans a safe, healthful, productive, and aesthetically and culturally pleasing environment. NEPA requires federal agencies to prepare a detailed statement on the environmental effects of proposed federal actions that significantly affect the quality of human health (Council on Environmental Quality, 1997).

The order calls for improved methodologies for assessing and mitigating impacts and health effects from multiple and cumulative exposure and collection of data on low-income and minority populations who may be disproportionately at risk and impacts on subsistence fishers and wildlife consumers. It also encourages participation of the affected populations in the various phases of assessing impacts, including scoping, data gathering, alternatives, analysis, mitigation, and monitoring.

The order focuses on "subsistence" fishers and wildlife consumers. Not everyone buys the fish they consume at the supermarket. There are many people who are subsistence fishers, who fish for protein, who basically subsidize their budgets, and their diets, by fishing from rivers, streams, and lakes that happen to be polluted. These subpopulations may be underprotected when basic assumptions are made using the dominant risk paradigm.

Communities Under Siege

Numerous studies reveal that low-income persons and people of color have borne greater health and environmental risk burdens than the society at large (Cooney, 1999; Goldman, 1992; Goldman & Fitton, 1994; Institute of Medicine, 1999; Mann, 1991). A recent study from the Institute of Medicine (1999) concluded that government, public health officials, and the medical and scientific communities need to place a higher value on the problems and concerns of environmental-justice communities. The study also confirmed what most affected communities have known for decades, that is, that minority and low-income communities are (1) exposed to higher levels of pollution than the rest of the nation and (2) experience certain diseases in greater number than the more affluent, White communities (Institute of Medicine, 1999).

Elevated public health risks have been found in some populations even when social class is held constant. For example, race has been found to be independent of class in the distribution of air pollution, contaminated fish consumption, location of municipal landfills and incinerators, abandoned toxic waste dumps, cleanup of Superfund sites, and lead poisoning in children (Agency for Toxic Substances and Disease Registry, 1988; Bryant & Mohai, 1992; Commission for Racial Justice, 1987; Goldman & Fitton, 1994; Lavelle & Coyle, 1992; Pirkle et al., 1994; Stretesky & Hogan, 1998; West et al., 1990).

Childhood lead poisoning is another preventable disease that has not been eradicated. Figures reported in the July 1994 *Journal of the American Medical Association* in the Third National Health and Nutrition Examination Survey (NHANES III) revealed that 1.7 million children (8.9% of children aged 1–5) are lead-poisoned, defined as having blood lead levels equal to or above 10 micrograms per deciliter. The NHANES III data found African American children to be lead-poisoned at more than twice the rate of White children at every income level (Pirkle et al., 1994). Over 28.4% of all low-income African American children were lead-poisoned, compared to 9.8% of low-income White children. During the time period between 1976 and 1991, the decrease in blood lead levels for African American and Mexican American children lagged far behind that of White children.

In 1992 in California, a coalition of environmental, social justice, and civil libertarian groups joined forces to challenge the way the state carried out its screening of poor children for blood lead levels. The Natural Resources Defense Council, the National Association for the Advancement of Colored People Legal Defense and

Education Fund, the American Civil Liberties Union, and the Legal Aid Society of Alameda County, California, won an out-of-court settlement worth $15 million to $20 million for a blood lead-testing program. The lawsuit, *Matthews v. Coye*, involved the failure of the state of California to conduct federally mandated testing for lead on some 557,000 poor children who receive Medicaid (B. L. Lee, 1992). This historic agreement triggered similar lawsuits and actions in several other states that failed to live up to the federal mandates.

Federal, state, and local policies and practices have contributed to residential segmentation and unhealthy living conditions in poor, working-class, and people-of-color communities (Bullard & Johnson, 1997). Several recent California cases bring this point to light (B. L. Lee, 1995). Disparate highway siting and mitigation plans were challenged by community residents, churches, and the NAACP Legal Defense and Education Fund in *Clean Air Alternative Coalition v. United States Department of Transportation* (N.D. Cal. C-93-0721-VRW), involving the reconstruction of the earthquake-damaged Cypress Freeway in West Oakland. The plaintiffs wanted the downed Cypress Freeway (which split their community in half) rebuilt farther away. Although the plaintiffs were not able to get their plan implemented, they did change the course of the freeway in their out-of-court settlement.

The NAACP Legal Defense and Education Fund filed an administrative complaint, *Mothers of East Los Angeles, El Sereno Neighborhood Action Committee, El Sereno Organizing Committee, et al. v. California Transportation Commission, et al.* (before the U.S. Department of Transportation and U.S. Housing and Urban Development), challenging the construction of the 4.5-mile extension of the Long Beach Freeway in East Los Angeles through El Sereno, Pasadena, and South Pasadena. The plaintiffs argued that the state agencies' proposed mitigation measures to address noise, air, and visual pollution discriminated against the mostly Latino El Sereno community. For example, all of the planned freeway in Pasadena and 80% in South Pasadena will be below ground level. On the other hand, most of the freeway in El Sereno will be above ground. White areas were favored over the mostly Latino El Sereno in allocation of covered freeway, historic preservation measures, and accommodation to local schools (Bullard & Johnson, 1997; B. L. Lee, 1995).

Los Angeles residents and the NAACP Legal Defense and Education Fund have also challenged the inequitable funding and operation of bus transportation used primarily by low-income and people-of-color residents. A class action lawsuit was filed on behalf of 350,000 low-income, people-of-color bus riders represented by the Labor/Community Strategy Center, the Bus Riders Union, Southern Christian Leadership Conference, Korean Immigrant Workers Advocates, and individual bus riders. In *Labor/Community Strategy Center v. Los Angeles Metropolitan Transportation Authority* (Cal. CV 94-5936 TJH Mcx), the plaintiffs argued that the Los Angeles Metropolitan Transit Authority (MTA) used federal funds to pursue a policy of raising costs to bus riders (who are mostly poor and people of color) and reducing quality of service in order to fund rail and other projects in predominately White suburban areas (Mann, 1996).

In the end, the Labor/Community Strategy Center and its allies successfully challenged transit racism in Los Angeles. The group was able to win major fare and bus pass concessions from the Los Angeles MTA. It also forced the Los Angeles MTA to spend $89 million on 278 new clean-compressed natural gas buses.

Many of the nation's environmental policies distribute costs in a regressive pattern while providing disproportionate benefits for Whites and individuals who fall at the upper end of the education and income scale. A 1992 study reported in the *National Law Journal* uncovered glaring inequities in the way the federal EPA enforces its laws. Lavelle and Coyle (1992) found the following:

> There is a racial divide in the way the U.S. government cleans up toxic waste sites and punishes polluters. White communities see faster action, better results and stiffer penalties than communities where blacks, Hispanics and other minorities live. This unequal protection often occurs whether the community is wealthy or poor. (pp. S1–S2)

The *National Law Journal* study reinforced what many grassroots activists have known for decades: All communities are not treated the same (Lavelle & Coyle, 1992). Communities that are located on the "wrong side of the tracks" are at greater risk from exposure to lead, pesticides (in the home and workplace), air pollution, toxic releases, water pollution, solid and hazardous waste, raw sewage, and pollution from industries (Goldman & Fitten, 1994).

Relocation From "Mount Dioxin"

Margaret Williams, a 73-year-old retired Pensacola, Florida, schoolteacher, led a 5-year campaign to get her community relocated from the environmental and health hazards posed by the nation's third largest Superfund site. The Escambia Wood Treating site was dubbed "Mount Dioxin" because of the 60-foot-high mound of contaminated soil dug up from the neighborhood. The L-shaped mound holds 255,000 cubic yards of soil contaminated with dioxin, one of the most dangerous compounds ever made (Olinger, 1996). Williams led Citizens Against Toxic Exposure (CATE), a neighborhood organization formed to get relocation, into battle with EPA officials, who first proposed to move only the 66 households most affected by the site (U.S. EPA, 1996). After prodding from CATE, EPA then added 35 more households, for a total cost of $7.54 million.

The original government plan called for some 257 households, including an apartment complex, to be left out. CATE refused to accept any relocation plan unless everyone was moved. The partial relocation was tantamount to partial justice. CATE took its campaign on the road to EPA's NEJAC and was successful in getting NEJAC's Waste Subcommittee to hold a Superfund relocation roundtable in Pensacola. At this meeting, CATE's total neighborhood relocation plan won the backing of more than 100 grassroots organizations. EPA nominated the Escambia Wood Treating Superfund site as the country's first pilot program to help the agency develop a nationally consistent relocation policy that would consider not only toxic levels but welfare issues such as property values, quality of life, health, and safety.

On October 3, 1996, EPA officials agreed to move all 358 households from the site at an estimated cost of $18 million. EPA officials deemed the mass relocation as "cost efficient" after city planners decided to redevelop the area for light industry rather than clean the site to residential standards (Escobedo, 1996; Washington Post, 1996). This decision marked the first time that an African American community had been relocated under EPA's Superfund program and was hailed as a landmark victory for environmental justice (Escobedo, 1996).

From Dumping in Dixie to Corporate Welfare

The southern United States has become a "sacrifice zone" for the rest of the nation's toxic waste (Schueler, 1992, p. 45). A colonial mentality exists in Dixie through which local government and big business take advantage of people who are both politically and economically powerless. The region is stuck with a unique legacy: the legacy of slavery, Jim Crow, and White resistance to equal justice for all. This legacy has also affected race relations and the region's ecology.

The South is characterized by "look-the-other-way environmental policies and giveaway tax breaks" and as a place where "political bosses encourage outsiders to buy the region's human and natural resources at bargain prices" (Schueler, 1992, pp. 46–47). Lax enforcement of environmental regulations has left the region's air, water, and land the most industry-befouled in the United States.

Toxic waste discharge and industrial pollution are correlated with poorer economic conditions. In 1992, the Institute for Southern Studies' "Green Index" ranked Louisiana 49th out of 50 states in overall environmental quality. Louisiana is not a rich state by any measure. It ranks 45th in the nation in spending on elementary and secondary education, for example.

Ascension Parish typifies the toxic "sacrifice zone" model. In two parish towns of Geismar and St. Gabriel, 18 petrochemical plants are crammed into a 9.5-square-mile area. In Geismar, Borden Chemicals has released harmful chemicals into the environment that are health hazardous to the local residents, including ethylene dichloride, vinyl chloride monomer, hydrogen chloride, and hydrochloric acid (Barlett & Steele, 1998, p. 72).

Borden Chemicals has a long track record of contaminating the air, land, and water in Geismar. In March 1997, the company paid a fine of $3.5 million—the single largest in Louisiana history—for storing hazardous waste, sludges, and solid wastes illegally; failing to install containment systems; burning hazardous waste without a permit; neglecting to report the release of hazardous chemicals into the air; contaminating groundwater beneath the plant site (thereby threatening an aquifer that provides drinking water for residents of Louisiana and Texas); and shipping toxic waste laced with mercury to South Africa without notifying the EPA, as required by law (Barlett & Steele, 1998).

Louisiana could actually improve its general welfare by enacting and enforcing regulations to protect the environment (Templet, 1995). However, Louisiana citizens subsidize corporate welfare with their health and the environment (Barlett & Steele, 1998). A growing body of evidence shows that environmental regulations do not kill jobs. On the contrary, the data indicate that "states with lower pollution levels and better environmental policies generally have more jobs, better socioeconomic conditions and are more attractive to new business" (Templet, 1995, p. 37). Nevertheless, some states subsidize polluting industries in the return for a few jobs (Barlett & Steele, 1998). States argue that tax breaks help create jobs. However, the few jobs that are created come at a high cost to Louisiana taxpayers and the environment.

Nowhere is the polluter-welfare scenario more prevalent than in Louisiana. Corporations routinely pollute the air, ground, and drinking water while being subsidized by tax breaks from the state. The state is a leader in doling out corporate welfare to polluters (see Table 1). In the 1990s, the state wiped off the books $3.1 billion in property taxes owed by polluting companies. The state's top five worst polluters received $111 million dollars over the past decade (Barlett & Steele, 1998). A breakdown of the chemical releases and tax breaks includes

- Cytec Industries (24.1 million pounds of releases/$19 million tax breaks)
- IMC-Agrico Co. (12.8 million pounds/$15 million)
- Rubicon, Inc. (8.4 million pounds/$20 million)
- Monsanto Co. (7.7 million pounds/$45 million)
- Angus Chemical Co. (6.3 million pounds/$12 million)

Not only is subsidizing polluters bad business, but it does not make environmental sense. For example, nearly three-fourths of Louisiana's population—more than 3 million people—get their drinking water from underground aquifers. Dozens of the aquifers are threatened by contamination from polluting industries (O'Byrne & Schleifstein, 1991). The Lower Mississippi River Industrial Corridor has over 125 companies that manufacture a range of products, including fertilizers, gasoline, paints, and plastics. This corridor has been dubbed "Cancer Alley" by environmentalists and local residents (Beasley, 1990a, 1990b; Bullard, 1994a; Motavalli, 1998).

Table 1. Corporate Welfare in Louisiana

The biggest recipients: Companies ranked by total industrial-property tax abatements, 1988–97

Company		Jobs created	Total taxes abated
1.	Exxon Corp.	305	$213,000,000
2.	Shell Chemical/Refining	167	$140,000,000
3.	International Paper	172	$103,000,000
4.	Dow Chemical Co.	9	$ 96,000,000
5.	Union Carbide	140	$ 53,000,000
6.	Boise Cascade Corp.	74	$ 53,000,000
7.	Georgia Pacific	200	$ 46,000,000
8.	Willamette Industries	384	$ 45,000,000
9.	Procter & Gamble	14	$ 44,000,000
10.	Westlake Petrochemical	150	$ 43,000,000

The costliest jobs: Companies ranked by net cost of each new job (abatements divided by jobs created).

Company		Jobs created	Cost per job
1.	Mobil Oil Corp.	1	$29,100,000
2.	Dow Chemical Co.	9	$10,700,000
3.	Olin Corp.	5	$ 6,300,000
4.	BP Exploration	8	$ 4,000,000
5.	Procter & Gamble	14	$ 3,100,000
6.	Murphy Oil USA	10	$ 1,600,000
7.	Star Enterprise	9	$ 1,500,000
8.	Cytec	13	$ 1,500,000
9.	Montell USA	31	$ 1,200,000
10.	Uniroyal Chemical Co.	22	$ 900,000

Note. Donald L. Bartlett and James B. Steele, "Paying a Price for Polluters," *Time*, November 23, 1998, p. 77. ©1998 by Time. Reprinted with permission.

Winning in Court: The Case of *CANT v. LES*

Executive Order 12898 was put to the test in rural Northwest Louisiana in 1989. Beginning that year, the Nuclear Regulatory Commission (NRC) had under review a proposal from Louisiana Energy Services (LES) to build the nation's first privately owned uranium enrichment plant. A national search was undertaken by LES to find the "best" site for a plant that would produce 17% of the nation's enriched uranium. LES supposedly used an objective scientific method in designing its site selection process.

The Southern United States, Louisiana, and Claiborne Parish ended up being the dubious "winners" of the site selection process. Residents from Homer and the nearby communities of Forest Grove and Center Springs—two communities closest to the proposed site—disagreed with the site selection process and outcome. They organized themselves into a group called Citizens Against Nuclear Trash (CANT), which charged LES and the NRC staff with practicing environmental racism. CANT hired the Sierra Club Legal Defense Fund (which later changed its name to Earthjustice Legal Defense Fund) and sued LES.

The lawsuit dragged on for more than 8 years. On May 1, 1997, a three-judge panel of the NRC's Atomic Safety and Licensing Board issued a final initial decision on the case. The judges concluded that "racial bias played a role in the selection process" (Nuclear Regulatory Commission, 1997). A story in the *London Sunday Times* proclaimed the environmental justice victory by declaring "Louisiana Blacks Win Nuclear War" (1997). The precedent-setting federal court ruling came 2 years after President Clinton signed Executive Order 12898. The judges, in a 38-page written decision, also chastised the NRC staff for not addressing the provision called for under Executive Order 12898. The court decision was upheld on appeal on April 4, 1998.

A clear racial pattern emerged during the so-called national search and multistage screening and selection process (Bullard, 1995). For example, as shown in Table 2, African Americans comprise about 13% of the U.S. population, 20% of the Southern states' population, 31% of Louisiana's population, 35% of Louisiana's northern parishes, and 46% of Claiborne Parish. This progressive narrowing of the site selection process to areas of increasingly high poverty and African American representation is also evident from an evaluation of the actual sites that were considered in the "intermediate" and "fine" screening stages of the site selection process. As noted in Table 3, the aggregate average percentage of Black population for a one-mile radius around all of the 78 sites examined (in 16 parishes) is

Table 2. Percentage of African American Population by Geographic Location, National Search for Privately Owned Uranium Enrichment Plant

Geographic location	Percentage African American (1990)
United States	13
Southern states	20
State of Louisiana	31
Louisiana's northern parishes	35
Claiborne Parish	46

Note. See U. S. Census Bureau, 1990 U.S. Census Data, PL 94-171 (visited May 10, 1999), http://www.census.gov

Table 3. Population by Race Living Within One-Mile Radius of LES Candidate Sites During Winnowing Process

Candidate sites	Total population	Black population	Percentage Black
Initial (78 sites)	18,722	5,321	28.35
Intermediate (37 sites)	8,380	3,082	36.78
Fine-screening (6 sites)	1,160	752	64.74
Final selection (1 site)	138	134	97.10

Note. See U. S. Census Bureau, 1990 U.S. Census Data, PL 94-171 (visited May 10, 1999), http://www.census.gov

28.35%. When LES completed its initial site cuts and reduced the list to 37 sites within nine parishes, the aggregate percentage of Black population rose to 36.78%. When LES then further limited its focus to six sites in Claiborne Parish, the aggregate average percentage Black population rose again, to 64.74%. The final site selected, the LeSage site, has a 97.10% Black population within a one-mile radius.

The LES plant was proposed to be built on Parish Road 39 between two African American communities—just one-quarter mile from Center Springs (founded in 1910) and one and one-quarter mile from Forest Grove (founded in the 1860s just after slavery). The proposed site is in a Louisiana parish that has per capita earnings of only $5,800 per year (just 45% of the national average of almost $12,800) and where over 58% of the African American population is below the poverty line. The two African American communities were rendered "invisible," since they were not even mentioned in the NRC's draft environmental impact statement (Nuclear Regulatory Commission, 1997).

Only after intense public comments did the NRC staff attempt to address environmental justice and disproportionate-impact implications, as required under the NEPA and called for under Executive Order 12898. For example, NEPA requires that the government consider the environmental impacts and weigh the costs and benefits of any proposed action. These include health and environmental effects, the risk of accidental but foreseeable adverse health and environmental effects, and socioeconomic impacts.

The NRC staff devoted less than a page to addressing environmental justice concerns of the proposed uranium enrichment plant in its final environmental impact statement (FEIS). Overall, the FEIS and Environmental Report (ER) are inadequate in the following respects: (1) they inaccurately assess the costs and benefits of the proposed plant, (2) they fail to consider the inequitable distribution of costs and benefits of the proposed plant to the White and African American population, and (3) they fail to consider the fact that the siting of the plant in a community of color follows a national pattern in which institutionally biased decision making leads to the siting of hazardous facilities in communities of color and results in the inequitable distribution of costs and benefits to those communities.

Among the distributive costs not analyzed in relationship to Forest Grove and Center Springs include the disproportionate burden of health and safety, diminished property values, fire and accidents, noise, traffic, radioactive dust in the air and water, and dislocation by closure of a road that connects the two communities. Overall, the CANT legal victory points to the utility of combining environmental and civil rights laws and the requirement of governmental agencies to consider Executive Order 12898 in their assessments.

In addition to the remarkable victory over LES, a company that had the backing of powerful U.S. and European nuclear energy companies, CANT members and their allies won much more. They empowered themselves and embarked on a path of political empowerment and self-determination. During the long battle,

CANT member Roy Mardris was elected to the Claiborne Parish Jury (i.e., county commission), and CANT member Almeter Willis was elected to the Claiborne Parish School Board. The town of Homer, the nearest incorporated town to Forest Grove and Center Springs, elected its first African American mayor, and the Homer town council now has two African American members. In fall 1998, LES sold the land on which the proposed uranium enrichment plant would have been located. The land is going back into timber production, for which it was used before LES bought it.

Winning on the Ground: *St. James Citizens v. Shintech*

Battle lines were drawn in Louisiana in 1991 in another national environmental justice test case. The community is Convent and the company is Shintech. The Japanese- owned Shintech, Inc., applied for a Title V air permit to build a $800 million polyvinyl chloride (PVC) plant in Convent, Louisiana, a community that is over 70% African American; over 40% of the Convent residents fall below the poverty line. The community already has a dozen polluting plants and yet has a 60% unemployment rate. The plants are so close to residents' homes, they could walk to work. The Black community is lured into accepting the industries with the promise of jobs, but in reality, the jobs are not there for local residents.

The Shintech case raised similar environmental racism concerns as those found in the failed LES siting proposal. The EPA is bound by Executive Order 12898 to ensure that "no segment of the population, regardless of race, color, national origin, or income, as a result of EPA's policies, programs, and activities, suffer disproportionately from adverse health or environmental effects, and all people live in clean and sustainable communities." The Louisiana Department of Environmental Quality is also bound by federal laws (e.g., Title VI of the Civil Rights Act of 1964) to administer and implement its programs, mandates, and policies in a nondiscriminatory way.

Any environmental justice analysis of the Shintech proposal will need to examine the issues of disproportionate and adverse impact on low-income and minority populations near the proposed PVC plant. Clearly, it is African Americans and low-income residents in Convent who live closest to existing and proposed industrial plants and who will be disproportionately impacted by industrial pollution (Wright, 1998). African Americans comprise 34% of the state's total population. The Shintech plant was planned for the St. James Parish, which ranks third in the state for toxic releases and transfers. Over 83% of St. James Parish's 4,526 residents are African American. Over 17.7 million pounds of releases were reported in the 1996 *Toxic Release Inventory (TRI)*. The Shintech plant would add over 600,000 pounds of air pollutants annually. Permitting the Shintech plant in Convent would add significantly to the toxic burden borne by residents, who are mostly low-income and African American.

After 6 months of intense organizing and legal maneuvering, residents of tiny Convent and their allies convinced EPA administrator Carol M. Browner to place the permit on hold. A feature article in *USA Today* bore the headline "EPA Puts Plant on Hold in Racism Case" (Hoversten, 1997). A year later, the Environmental Justice Coalition forced Shintech to scrap its plans to build the PVC plant in the mostly African American community. The decision came in September 1998 and was hailed around the country as a major victory against environmental racism. The driving force behind this victory was the relentless pressure and laser-like focus of the local Convent community.

Radioactive Colonialism and Native Lands

There is a direct correlation between exploitation of land and exploitation of people. It should not be a surprise to anyone to discover that Native Americans have to contend with some of the worst pollution in the United States (Beasley, 1990b; Kay, 1991; Taliman, 1992; Tomsho, 1990). Native American nations have become prime targets for waste trading (Angel, 1992; Geddicks, 1993). More than three dozen Indian reservations have been targeted for landfills, incinerators, and other waste facilities (Kay, 1991). The vast majority of these waste proposals have been defeated by grassroots groups on the reservations. However, "radioactive colonialism" is alive and well (Churchill & LaDuke, 1983).

Radioactive colonialism operates in energy production (mining of uranium) and disposal of wastes on Indian lands. The legacy of institutional racism has left many sovereign Indian nations without an economic infrastructure to address poverty, unemployment, inadequate education and health care, and a host of other social problems.

Some industry and governmental agencies have exploited the economic vulnerability of Indian nations. For example, of the 21 applicants for the DOE's monitored retrievable storage (MRS) grants, 16 were Indian tribes (Taliman, 1992a). The 16 tribes lined up for $100,000 grants from the DOE to study the prospect of "temporarily" storing nuclear waste for a half century under its MRS program.

It is the Native American tribes' sovereign right to bid for the MRS proposals and other industries. However, there are clear ethical issues involved when the U.S. government contracts with Indian nations that lack the infrastructure to handle dangerous wastes in a safe and environmentally sound manner. Delegates at the Third Annual Indigenous Environmental Council Network Gathering (held in Cello Village, Oregon, on June 6, 1992) adopted a resolution of "No nuclear waste on Indian lands."

Transboundary Waste Trade

Hazardous waste generation and international movement of hazardous waste pose some important health, environmental, legal, and ethical dilemmas. It is unlikely that many of the global hazardous waste proposals can be effectuated without first addressing the social, economic, and political context in which hazardous wastes are produced (industrial processes), controlled (regulations, notification and consent documentation), and managed (minimization, treatment, storage, recycling, transboundary shipment, pollution prevention).

The "unwritten" policy of targeting Third World nations for waste trade received international media attention in 1991. Lawrence Summers, at the time he was chief economist of the World Bank, shocked the world and touched off an international scandal when his confidential memorandum on waste trade was leaked. Summers writes: "'Dirty' Industries: Just between you and me, shouldn't the World Bank be encouraging *more* migration of the dirty industries to the LDCs?" (Quoted in Greenpeace, 1993, pp. 1–2).

Consumption and production patterns, especially in nations with wasteful "throw-away" lifestyles like the United States, and the interests of transnational corporations create and maintain unequal and unjust waste burdens within and between affluent and poor communities, states, and regions of the world. Shipping hazardous wastes from rich communities to poor communities is not a solution to the growing global waste problem. Not only is it immoral, but it should be illegal. Moreover, making hazardous waste transactions legal does not address the ethical issues imbedded in such transactions (Alston & Brown, 1993).

Transboundary shipment of banned pesticides, hazardous wastes, toxic products, and export of "risky technologies" from the United States, where regulations and laws are more stringent, to nations with weaker infrastructure, regulations, and laws smacks of a double standard (Bright, 1990). The practice is a manifestation of power arrangements and a larger stratification system in which some people and some places are assigned greater value than others.

In the real world, all people, communities, and nations are *not* created equal. Some populations and interests are more equal than others. Unequal interests and power arrangements have allowed poisons of the rich to be offered as short-term remedies for poverty of the poor. This scenario plays out domestically (as in the United States, where low-income and people of color communities are disproportionately affected by waste facilities and "dirty" industries) and internationally (where hazardous wastes move from OECD states flow to non-OECD states).

The conditions surrounding the more than 1,900 maquiladoras (assembly plants operated by American, Japanese, and other foreign countries) located along the 2,000-mile U.S.-Mexico border may further exacerbate the waste trade (Sanchez, 1990). The maquiladoras use cheap Mexican labor to assemble imported components and raw material and then ship finished products back to the United States. Nearly a half million Mexican workers are employed in the maquiladoras.

A 1983 agreement between the United States and Mexico required American companies in Mexico to return their waste products to the United States. Plants were required to notify the U.S. EPA when returning wastes. Results from a 1986 survey of 772 maquiladoras revealed that only 20 of the plants informed the EPA that they were returning waste to the United States, even though 86% of the plants used toxic chemicals in their manufacturing process (Juffers, 1988). In 1989, only 10 waste shipment notices were filed with the EPA (Center for Investigative Reporting, 1990).

Much of the wastes end up being illegally dumped in sewers, ditches, and the desert. All along the Lower Rio Grande River Valley, maquiladoras dump their toxic wastes into the river, from which 95% of the region's residents get their drinking water (Hernandez, 1993). In the border cities of Brownsville, Texas, and Matamoras, Mexico, the rate of anencephaly—babies born without brains—is four times the national average. Affected families have filed lawsuits against 88 of the area's 100 maquiladoras for exposing the community to xylene, a cleaning solvent that can cause brain hemorrhages and lung and kidney damage.

The Mexican environmental regulatory agency is understaffed and ill-equipped to adequately enforce its laws (Barry & Simms, 1994; Working Group on Canada-Mexico Free Trade, 1991). Only time will tell if the North American Free Trade Agreement (NAFTA) will "fix" or exacerbate the public health and the environmental problems along the U.S.-Mexico border.

Conclusion

The environmental protection apparatus in the United States does not provide equal protection for all communities. The environmental justice movement emerged in response to environmental inequities, threats to public health, unequal protection, differential enforcement, and disparate treatment received by the poor and people of color. The movement redefined environmental protection as a basic right. It also emphasized pollution prevention, waste minimization, and cleaner production techniques as strategies for achieving environmental justice for all Americans without regard to race, color, national origin, or income.

The poisoning of African Americans in Louisiana's "Cancer Alley," Native Americans on reservations, and Mexicans in the border towns all have their roots in the same economic system, a system characterized by economic exploitation, racial oppression, and devaluation of human life and the natural environment. Both race and class factors place low-income and people-of-color communities at special risk. Although environmental and civil rights laws have been on the books for more than 3 decades, all communities have not received the same benefits from their application, implementation, and enforcement.

Unequal political power arrangements also have allowed poisons of the rich to be offered as short-term economic remedies for poverty. There is little or no correlation between proximity of industrial plants in communities of color and the employment opportunities of nearby residents. Having industrial facilities in one's community does not automatically translate into jobs for nearby residents. Many industrial plants are located at the fence line with the communities. Some are so close that local residents could walk to work. More often than not, communities of color are stuck with the pollution and poverty, while other people commute in for the industrial jobs.

Similarly, tax breaks and corporate welfare programs have produced few new jobs by polluting firms. However, state-sponsored pollution and lax enforcement have allowed many communities of color and poor communities to become the dumping grounds. Louisiana is the poster child for corporate welfare. The state is mired in both poverty and pollution. It is no wonder that Louisiana's petrochemical corridor, the 85-mile stretch along the Mississippi River from Baton Rouge to New Orleans dubbed "Cancer Alley," has become a hotbed for environmental justice activity.

The environmental justice movement has set out clear goals of eliminating unequal enforcement of environmental, civil rights, and public health laws; differential exposure of some populations to harmful chemicals, pesticides, and other toxins in the home, school, neighborhood, and workplace; faulty assumptions in calculating, assessing, and managing risks; discriminatory zoning and land use practices; and exclusionary policies and practices that limit some individuals and groups from participation in decision making. Many of these problems could be eliminated if existing environmental, health, housing, and civil rights laws were vigorously enforced in a nondiscriminatory way.

The call for environmental and economic justice does not stop at the U.S. borders but extends to communities and nations that are threatened by the export of hazardous wastes, toxic products, and "dirty" industries. Much of the world does not get to share in the benefits of the United States' high standard of living. From energy consumption to the production and export of tobacco, pesticides, and other chemicals, more and more of the world's peoples are sharing the health and environmental burden of America's wasteful throwaway culture. Hazardous wastes and "dirty" industries have followed the path of least resistance. Poor people and poor nations are given a false choice of "no jobs and no development" versus "risky, low-paying jobs and pollution."

Industries and governments (including the military) have often exploited the economic vulnerability of poor communities, poor states, poor nations, and poor regions for their unsound and "risky" operations. Environmental justice leaders are demanding that no community or nation, rich or poor, urban or suburban, Black or White, be allowed to become a "sacrifice zone" or dumping grounds. They are also pressing governments to live up to their mandate of protecting public health and the environment.

References

Agency for Toxic Substances and Disease Registry. (1988). *The nature and extent of lead poisoning in children in the United States: A report to Congress.* Atlanta, GA: U.S. Department of Health and Human Services.

Alston, D. (1992). Transforming a movement: People of color unite at summit against environmental racism. *Sojourner, 21*(1), 30–31.

Alston, D., & Brown, N. (1993). Global threats to people of color. In R. D. Bullard (Ed.), *Confronting environmental racism: Voices from the grassroots* (pp. 179–194). Boston: South End Press.

Angel, B. (1992). *The toxic threat to Indian lands: A Greenpeace report.* San Francisco: Greenpeace.

Austin, R., & Schill, M. (1991). Black, Brown, poor, and poisoned: Minority grassroots environmentalism and the quest for eco-justice. *Kansas Journal of Law and Public Policy, 1*(1), 69–82.

Barlett, D. L., & Steele, J. B. (1998, November 23). Paying a price for polluters. *Time,* pp. 72–80.

Barry, T., & Simms, B. (1994). *The challenge of cross border environmentalism: The U.S.-Mexico case.* Albuquerque, NM: Inter-Hemispheric Education Resource Center.

Beasley, C. (1990a). Of pollution and poverty: Keeping watch in Cancer Alley. *Buzzworm, 2*(4), 39–45.

Beasley, C. (1990b). Of poverty and pollution: Deadly threat on native lands. *Buzzworm, 2*(5), 39–45.

Bryant, B. (1995). *Environmental justice: Issues, policies, and solutions.* Washington, D.C.: Island Press.

Bryant, B., & Mohai, P. (1992). *Race and the incidence of environmental hazards.* Boulder, CO: Westview Press.

Bullard, R. D. (1983). Solid waste sites and the Black Houston community. *Sociological Inquiry, 53* (Spring): 273–288.

Bullard, R. D. (1990). Ecological inequities and the New South: Black communities under siege. *Journal of Ethnic Studies, 17* (Winter), 101–115.

Bullard, R. D. (1992). *The environmental justice framework: A stategy for addressing unequal protection.* Paper presented at the Resources for the Future Conference on Risk Management, Annapolis, MD.

Bullard, R. D. (1993a). *Confronting environmental racism: Voices from the grassroots.* Boston: South End Press.

Bullard, R. D. (1993b). Race and environmental justice in the United States. *Yale Journal of International Law, 18*(1), 319–355.

Bullard, R. D. (1993c). Environmental racism and land use. *Land Use Forum: A Journal of Law, Policy & Practice, 2*(1), 6–11.

Bullard, R. D. (1994a). *Dumping in Dixie: Race, class and environmental quality.* Boulder, CO: Westview Press.

Bullard, R. D. (1994b). Grassroots flowering: The environmental justice movement comes of age. *Amicus, 16* (Spring), 32–37.

Bullard, R. D. (1995). Prefiled written testimony at the *CANT v. LES* hearing, Shreveport, LA.

Bullard, R. D. (1996). *Unequal protection: Environmental justice and communities of color.* San Francisco: Sierra Club.

Bullard, R. D., & Feagin, J. R. (1991). Racism and the city. In M. Gottdiener & C. V. Pickvance (Eds.), *Urban life in transition* (pp. 55–76). Newbury Park, CA: Sage.

Bullard, R. D., & Johnson, G. S. (1997). *Just transportation: Dismantling race and class barriers to mobility.* Gabriola Island, British Columbia, Canada: New Society.

Calloway, C. A., & Decker, J. A. (1997). Environmental justice in the United States: A primer. *Michigan Bar Journal, 76* (January), 62–68.

Center for Investigative Reporting. (1990). *Global dumping grounds: The international traffic in hazardous waste.* Washington, DC: Seven Locks Press.

Chase, A. (1993). Assessing and addressing problems posed by environmental racism. *Rutgers University Law Review, 45*(2), 385–369.

Churchill, W., & LaDuke, W. (1983). Native America: The political economy of radioactive colonialism. *Insurgent Sociologist, 13*(1), 51–63.

Coleman, L.A. (1993). It's the thought that counts: The intent requirement in environmental racism claims. *St. Mary's Law Journal, 25*(1), 447–492.

Collin, R. W. (1992). Environmental equity: A law and planning approach to environmental racism. *Virginia Environmental Law Journal, 13*(4), 495–546.

Collin, R. W., & Collin, R. M. (1998). The role of communities in environmental decisions: Communities speaking for themselves. *Journal of Environmental Law and Litigation, 13*, 3789.

Colopy, J. H. (1994). The road less traveled: Pursuing environmental justice through Title VI of the Civil Rights Act of 1964. *Stanford Environmental Law Journal, 13*(1), 125–189.

Colquette, K. C., & Robertson, E. A. H. (1991). Environmental racism: The causes, consequences, and commendations. *Tulane Environmental Law Journal, 5*(1), 153–207.

Commission for Racial Justice. (1987). *Toxic wastes and race in the United States*. New York: United Church of Christ.

Cooney, C. M. (1999). Still searching for environmental justice. *Environmental Science and Technology, 33* (May), 200–205.

Council on Environmental Quality. (1971). *Second annual report of the Council on Environmental Quality*. Washington, DC: U.S. Government Printing Office.

Council on Environmental Quality. (1997). *Environmental justice: Guidance under the National Environmental Policy Act*. Washington, DC: Author.

Doob, C. B. (1993). *Racism: An American cauldron*. New York: Harper Collins.

Edelstein, M. R. (1990). Global dumping grounds. *Multinational Monitor, 11*(1), 26–29.

Escobedo, D. (1996, October 4). EPA gives in, will move all at toxic site. *Pensacola News Journal*, p. A1.

Faber, D. (1998). *The struggle for environmental justice*. New York: Guilford Press.

Geddicks, A. (1993). *The new resource wars: Native and environmental struggles against multinational corporations*. Boston: South End Press.

Godsil, R. D. (1990). Remedying environmental racism. *Michigan Law Review, 90*, 394–427.

Goldman, B. (1992). *The truth about where you live: An atlas for action on toxins and mortality*. New York: Random House.

Goldman, B., & Fitten, L. J. (1994). *Toxic wastes and race revisited*. Washington, DC: Center for Policy Alternatives/National Association for the Advancement of Colored People/United Church of Christ.

Greenpeace. (1990). *The international trade in waste: A Greenpeace inventory*. Washington, DC: Greenpeace USA.

Greenpeace. (1993). *The case for a ban on all hazardous waste shipment from the United States and other OECD member states to non-OECD states*. Washington, DC: Greenpeace USA.

Hernandez, B. J. (1993). Dirty growth. *New Internationalist* (August).

Hoversten, P. (1997, September 11). EPA puts plant on hold in racism case. *USA Today*, p. A3.

Institute for Southern Studies. (1992). *1991–1992 Green Index: A state-by-state guide to the nation's environmental health*. Durham, NC: Author.

Institute of Medicine. (1999). *Toward environmental justice: Research, education, and health policy needs*. Washington, DC: National Academy Press.

Johnson, B. L., Williams, R. C., & Harris, C. M. (Eds.). (1992). *Proceedings of the 1990 National Minority Health Conference: Focus on Environmental Contamination*. Princeton, NJ: Scientific.

Juffers, J. (1988, October 24). Dump at the border: U.S. firms make a Mexican wasteland. *Progressive*.

Kay, J. (1991, April 10). Indian lands targeted for waste disposal sites. *San Fransciso Examiner*.

Lavelle, M., & Coyle, M. (1992). Unequal protection. *National Law Journal*, 1–2.

Lee, B. L. (1992, February). *Environmental litigation on behalf of poor, minority children*: Matthews v. Coye: *A case study*. Paper presented at the annual meeting of the American Association for the Advancement of Science, Chicago.

Lee, B. L. (1995, May). *Civil rights remedies for environmental injustice*. Paper presented at Transportation and Environmental Justice: Building Model Partnerships conference, Atlanta, GA.

Lee, C. (1992). *Proceedings: The First National People of Color Environmental Leadership Summit*. New York: United Church of Christ, Commission for Racial Justice.

Leonard, A. (1992). Plastics: Trashing the third world. *Multinational Monitor* (June), 26–31.

Leonard, A. (1993). Poison fields: Dumping toxic "fertilizer" on Bangladeshi farmers. *Multinational Monitor* (April), 14–18.

Livingston, M. L. (1989). Transboundary environmental degradation: Market failure, power, and instrumental justice. *Journal of Economic Issues, 23*, 79–91.

Louisiana Blacks win nuclear war. (1997, May 11). *Sunday London Times*.

Mann, E. (1991). *L.A.'s lethal air: New strategies for policy, organizing, and action*. Los Angeles: Labor/Community Strategy Center.

Mann, E. (1996). *A new vision for urban transportation: The bus riders union makes history at the intersection of mass transit, civil rights, and the environment*. Los Angeles: Labor/Community Strategy Center.

Motavalli, J. (1998). Toxic targets: Polluters that dump on communities of color are finally being brought to justice. *E* (July/August), 28–41.

National Institute for Environmental Health Sciences. (1995). *Proceedings of the Health and Research Needs to Ensure Environmental Justice Symposium*. Research Triangle Park, NC: Author.

Nieves, L. A. (1992, February). *Not in whose backyard? Minority population concentrations and noxious facility sites*. Paper presented at the annual meeting of the American Association for the Advancement of Science, Chicago.

Nuclear Regulatory Commission. (1997). *Final initial decision—Louisiana Energy Services*. U.S. Nuclear Regulatory Commission, Atomic Safety and Licensing Board, Docket no. 70-3070-ML. May 1.

O'Byrne, & Schleifstein, M. (1991, February 19). Drinking water in danger. *Times Picayune*, p. A5.

Pirkle, J. L., Brody, D. J., Gunter, E. W., Kramer, R. A., Paschal, D. C., Flegal, K. M., & Matte, T. D. (1994). The decline in blood lead levels in the United States: The National Health and Nutrition Examination Survey (NHANES III). *Journal of the American Medical Association, 272*, 284–291.

Puckett, J. (1993). *Basel: Another dumping convention*. Washington, DC: Greenpeace.

Russell, D. (1989). Environmental racism. *Amicus, 11*(2), 22–32.

Sanchez, R. (1990). Health and environmental risks of the maquiladora in Mexicali. *National Resources Journal, 30*(1), 163–186.

Schueler, D. (1992). Southern exposure. *Sierra, 77* (November–December), 45–47.

Stretesky, P., & Hogan, M. J. (1998). Environmental justice: An analysis of Superfund sites in Florida. *Social Problems, 45* (May), 268–287.

Taliman, V. (1992a). Stuck holding the nation's nuclear waste. *Race, Poverty, & Environment Newsletter* (Fall), 6–9.

Taliman, V. (1992b). The toxic waste of Indian lives. *Covert Action, 40*(1), 16–19.

Templet, P. T. (1995). The positive relationship between jobs, environment and the economy: An empirical analysis and review. *Spectrum* (Spring), 37–49.

Tomsho, R. (1990, November 29). Dumping grounds: Indian tribes contend with some of the worst of America's pollution. *Wall Street Journal*.

United Nations. (1993). Environmentally sound management of hazardous wastes, including prevention of illegal international traffic in hazardous wastes. In *United Nations Summit on Environment and Development: Agenda 21* (pp. 197–205). New York: Author.

U.S. Environmental Protection Agency. (1991). *Hazardous waste exports by receiving country*. Washington, DC: Author.

U.S. Environmental Protection Agency. (1992a). *Environmental equity: Reducing risk for all communities*. Washington, DC: Author.

U.S. Environmental Protection Agency. (1992b). Geographic initiatives: Protecting what we love. *Securing our legacy: An EPA progress report 1989–1991*. Washington, DC: Author.

U.S. Environmental Protection Agency. (1993). *Toxic release inventory and emission reductions 1987–1990 in the Lower Mississippi River industrial corridor*. Washington, DC: EPA, Office of Pollution Prevention and Toxics.

U.S. Environmental Protection Agency. (1996). *Escambia Treating Company interim action: Addendum to April 1996 Superfund proposed plan fact sheet*. Atlanta, GA: EPA, Region IV.

U.S. Environmental Protection Agency. (1998). *Guidance for incorporating environmental justice in EPA's NEPA compliance analysis*. Washington, DC: Author.

U.S. General Accounting Office. (1983). *Siting of hazardous waste landfills and their correlations with racial and economic status of surrounding communities*. Washington, DC: U.S. Government Printing Office.

U.S. to move families away from Florida toxic dump. *Washington Post*, October 6, 1996.

Wernette, D. R., & Nieves, L. A. (1992). Breathing polluted air. *EPA Journal, 18*(1), 16–17.

West, P., Fly, J. M., Larkin, F., & Marans, P. (1990). Minority anglers and toxic fish consumption: Evidence of the state-wide survey of Michigan. In B. Bryant & P. Mohai (Eds.), *Race and the incidence of environmental hazards* (pp. 100–113). Boulder, CO: Westview.

Wright, B. H. (1998). *St. James Parish field observations*. New Orleans, LA: Xavier University, Deep South Center for Environmental Justice.

ROBERT D. BULLARD is the Ware Professor of Sociology and Director of the Environmental Justice Resource Center at Clark Atlanta University. His book *Dumping in Dixie: Race, Class and Environmental Quality* (Westview Press, 1994) is a standard text in the environmental justice field. A few of his other related books include *Confronting Environmental Racism: Voices from the Grassroots* (South End Press, 1993) and *Unequal Protection: Environmental Justice and Communities of Color* (Sierra Club Books, 1996). His most recent book, coedited with Glenn S. Johnson, is entitled *Just Transportation: Dismantling Race and Class Barriers to Mobility* (New Society Publishers, 1997). He is completing work on a new book on urban sprawl to be published by Island Press.

GLENN S. JOHNSON is a Research Associate in the Environmental Justice Resource Center and an Assistant Professor in the Department of Sociology at Clark Atlanta University. He is coeditor, with Robert D. Bullard, of *Just Transportation: Dismantling Race and Class Barriers to Mobility* (New Society Publishers, 1997) and coeditor of a special-edition, two-issue series of *Race, Gender and Class*, "Environmentalism and Race, Gender, Class Issues" (Part I–December 1997 and Part II–September 1998).

SCIENCE AND CULTURAL THEORY

A New Series Edited by Barbara Herrnstein Smith and E. Roy Weintraub

Evolution's Eye

A Systems View of the Biology-Culture Divide

SUSAN OYAMA

"Oyama writes elegantly and from a deep intellectual base. This alternative view to the dominant genetic determinism will be of interest to all who seek a more complex view of human nature. It is an excellent book, beautifully composed." —Katherine Nelson, City University of New York

"To think of nature and nurture as two distinct categories is not only wrong, Susan Oyama convincingly argues, but doing so hobbles our attempts to understand the nature of development and evolution at every level. Hers is a voice that needs to be heard."—Evelyn Fox Keller, Massachusetts Institute of Technology

288 pages, paper $18.95 *Science and Cultural Theory*

The Ontogeny of Information

Developmental Systems and Evolution

Second edition Revised and Expanded

SUSAN OYAMA

"The publication of this revised edition is timely and welcome, especially given the current dominance of simplistic views about genetic causation, aided by constant misuse of the ideas of information, coding and programming. Oyama's classic discussion of these concepts combines patient, subtle dissection with bold and novel moves. *The Ontogeny of Information* is a work of brilliant originality and enduring relevance."—Peter Godfrey-Smith, Stanford University

"This is among the most important books on developmental theory published in the last several decades."—Robert Lickliter, Virginia Polytechnic Institute

296 pages, paper $19.95 *Science and Cultural Theory*

TWO POSITIONS: ORGANIZATIONAL BEHAVIOR AND DEVELOPMENTAL PSYCHOLOGY – CLAREMONT GRADUATE UNIVERSITY

Claremont Graduate University announces two searches for tenure-track faculty (rank open) in (1) Organizational Behavior and (2) Developmental Psychology to teach and supervise research in the Ph.D. program in their area of specialization and to contribute to other academic programs, especially program evaluation; interdisciplinary collaboration is encouraged. Further information on these positions and other faculty positions available at CGU can be found under Employment Opportunities at http://www.cgu.edu/sbos